Regeneration Potential in Plants

Edited by
Dhananjay Kumar Pandey and
Bhupendra Chaudhary

CRC Press
Taylor & Francis Group
Boca Raton London New York

CRC Press is an imprint of the
Taylor & Francis Group, an **informa** business

Designed cover images: Shutterstock/Author's own

First edition published 2025
by CRC Press
2385 NW Executive Center Drive, Suite 320, Boca Raton FL 33431

and by CRC Press
4 Park Square, Milton Park, Abingdon, Oxon, OX14 4RN

CRC Press is an imprint of Taylor & Francis Group, LLC

Library of Congress Cataloging-in-Publication Data
Names: Pandey, Dhananjay Kumar, editor. | Chaudhary, Bhupendra, editor.
Title: Regeneration potential in plants / edited by Dhananjay Kumar Pandey
and Bhupendra Chaudhary.
Description: First edition | Boca Raton, FL : CRC Press, 2025. |
Includes bibliographical references and index.
Identifiers: LCCN 2024057695 (print) | LCCN 2024057696 (ebook) |
ISBN 9781032702407 (hardback) | ISBN 9781032702384 (paperback) |
ISBN 9781032702391 (ebook)
Subjects: LCSH: Regeneration (Botany) | Plant genetics. | Plant physiology.
Classification: LCC QK840 .R44 2025 (print) | LCC QK840 (ebook) |
DDC 571.8/89—dc23/eng/20250219
LC record available at https://lccn.loc.gov/2024057695
LC ebook record available at https://lccn.loc.gov/2024057696

ISBN: 9781032702407 (hbk)
ISBN: 9781032702384 (pbk)
ISBN: 9781032702391 (ebk)

DOI: 10.1201/9781032702391

Typeset in Times
by codeMantra

Contents

About the Editors

Dhananjay Kumar Pandey, PhD, is an accomplished academician and researcher with expertise in plant molecular biology and agronomic trait enhancement in crops. He currently serves as an assistant professor at Amity University, Jharkhand. With over eight years of research and teaching experience in India and abroad, he has made notable contributions to his field. His academic journey includes postdoctoral research stints at Adam Mickiewicz University, Poland, and the University of California, San Diego, USA. His research focuses on unraveling molecular mechanisms underlying plant development for improved crop traits. He has authored more than 20 papers in esteemed international journals and filed eight patents stemming from his pioneering research. Some key contributions include unraveling the transcriptional network of cotton fiber elongation, exploring the evolutionary diversity of plant profilin genes, and investigating novel strategies for crop biofortification. His research has global relevance, evident from collaborations across borders. He has also shared his insights through book chapters, notably in the fields of agri/bioinformatics and next-generation sequencing applications in biological sciences. As an assistant professor at Amity University, he continues to be a driving force in plant molecular biology research, offering valuable contributions to the advancement of agriculture and biotechnology.

Bhupendra Chaudhary, PhD, is an associate professor at the School of Life Sciences, Jawaharlal Nehru University in New Delhi, India. His expertise lies in genomics and genetic engineering, particularly applied to improving crop plants' attributes, stress resilience, and field traits, with a focus on cotton fibers. He has held leadership positions, such as being the Head of the Department at the School of Biotechnology at Gautam Buddha University. His academic journey includes earning his MPhil and PhD degrees from the University of Delhi in 1999 and 2006, respectively. He also contributed to international academia as a DBT visiting associate at the Department of Ecology, Evolution, and Organismal Biology at Iowa State University, USA, from 2007 to 2008. Prior to that, he served as a lecturer of botany at Visva-Bharati (Central) University, Santiniketan, between 2005 and 2009. His research dedication is reflected in his significant contribution to the field, with over 40 research papers published in respected international peer-reviewed journals. Additionally, he has successfully completed seven extramural research projects as the principal investigator, securing funding from various national agencies. Overall, his profile showcases a strong commitment to research and academic excellence in the field of genomics, genetic engineering, and crop enhancement.

Contributors

Akash Ajay
Department of Biotechnology
Vignan's Foundation for Science, Technology and Research
Guntur, India

Veluru Aparna
Division of Crop Improvement
ICAR-Central Plantation Crops Research Institute
Kasaragod, India

Sakshi Arora
Amity Institute of Microbial Technology
Amity University
Noida, India

Vedpriya Arya
Herbal Research Department
Patanjali Research Institute
Haridwar, India
University of Patanjali
Haridwar, India
Centre of Excellence
Patanjali Ayurveda Hospital
Haridwar, India

Ayushi
School of Life Sciences, JNU
Delhi, India

Acharya Balkrishna
Herbal Research Department
Patanjali Research Institute
Haridwar, India
University of Patanjali
Haridwar, India

Supriyo Basak
Amity Institute of Biotechnology
Amity University Jharkhand
Ranchi, India

Thirupataiah Boini
National Ayurveda Research Institute for Panchakarma
CCRAS, Ministry of Ayush
Thrissur, India

Bhupendra Chaudhary
School of Life Science, JNU
Delhi, India

Priyasha Das
Amity Institute of Microbial Technology
Amity University
Noida, India

Akash Deep
Amity Institute of Biotechnology
Amity University Jharkhand
Ranchi, India

Mahwish Firdous
Amity Institute of Biotechnology
Amity University Jharkhand
Ranchi, India

Dhananjay Suresh Gaikwad
Amity Institute of Organic Agriculture
Amity University
Noida, India

Parth Gharat
Regional Ayurveda Research Institute
CCRAS, Ministry of Ayush
Pune, India

Parth Ramesh Gharat
Regional Ayurveda Research Institute
CCRAS
Pune, India

Aakriti Gupta
Harlal Institute of Management and Technology
Noida, India

Pranay Guru
Bansal Group of Institute, RGPV
Bhopal, India

Prem Prakash Jangid
Regional Ayurveda Research Institute
CCRAS
Pune, India

Motha Kalpana
Dr. Y. S. R Horticultural University
Venkataramannagudem, India

Devinder Kaur
School of Life Sciences, JNU
Delhi, India

Anurag Kumar
Regional Ayurveda Research Institute
CCRAS
Pune, India

Brijesh Kumar
Herbal Research Department
Patanjali Research Institute
Haridwar, India

Kunal Kumar
Amity Institute of Biotechnology
Amity University Jharkhand
Ranchi, India

Rahul Maurya
National Ayurveda Research Institute for Panchakarma
Central Ayurveda Research Institute,
CCRAS, Ministry of Ayush
Thrissur, India

Lakshminarayana Misro
National Ayurveda Research Institute for Panchakarma
CCRAS, Ministry of Ayush
Thrissur, India

Dhananjay Kumar Pandey
Amity Institute of Biotechnology
Amity University Jharkhand
Ranchi, India

Kunal Ranjan
Amity Institute of Biotechnology
Amity University Jharkhand
Ranchi, India

Aakanksha Saini
Herbal Research Department
Patanjali Research Institute
Haridwar, India

Abha Sharma
Amity Institute of Organic Agriculture
Amity University
Noida, India

Rajani Sharma
Amity Institute of Biotechnology
Amity University Jharkhand
Ranchi, India

Ajay Kumar Shukla
Dr. Ram Manohar Lohia Avadh University
Ayodhya, India

Pathakakula Srinivasu
Tamil Nadu Agricultural University
Coimbatore, India

Sky
Amity Institute of Organic Agriculture
Amity University
Noida, India

Deepshikha Thakur
Amity Institute of Organic Agriculture
Amity University
Noida, India

Poonam Kumari Yadav
Amity Institute of Biotechnology
Amity University Jharkhand
Ranchi, India

Priya Yadav
Herbal Research Department
Patanjali Research Institute
Haridwar, India

Priyanka Yadav
Herbal Research Department
Patanjali Research Institute
Haridwar, India

1 Introduction to Plant Regeneration Concept and Significance

Rahul Maurya, Thirupataiah Boini, Lakshminarayana Misro, Parth Gharat, Ajay Kumar Shukla, and Pranay Guru

1.1 INTRODUCTION

1.1.1 Definition of Plant Regeneration

Plant regeneration refers to the ability of plants to regrow or develop new plant structures from various plant tissues, often in response to damage, stress, or as part of their everyday life cycle. This process allows plants to replace lost or damaged parts and is crucial for survival and adaptation (Bidabadi and Siamak, 2020). There are several mechanisms of plant regeneration, and examples include:

1. *Vegetative Propagation*: Some plants, such as strawberries, produce above-ground stems called runners or stolons that develop into new plants. However, Plants like bamboo and ginger produce underground stems called rhizomes, which can give rise to new shoots and roots (Guo and Madison, 2021).
2. *Budding*: Plants have buds in the leaf axils, which can develop into new shoots when activated. For example, roses can be propagated through cutting and planting axillary buds (Bredmose and Hansen, 1995).
3. *Cutting and Grafting*: Many plants can be propagated by taking cuttings from stems or leaves and planting them to generate new plants. The propagation of succulents or herbs like mint. However, Grafting involves joining the stem or bud of one plant (the scion) with the root system of another plant (the rootstock). Fruit trees like apples and grapevines are often propagated through grafting. This method also propagates baobab (*Adansonia digitata L.*) in Kenya to promote cultivation (Anjarwalla et al., 2017).
4. *Tissue Culture*: In vitro techniques like tissue culture involve growing plant cells, tissues, or organs in a controlled environment. This method regenerates whole plants from small parts, like callus or explants (Bansal and Manoj, 2024).
5. *Regeneration from Meristematic Cells*: The apical meristems at the tips of shoots and roots contain undifferentiated cells that can give rise to new

DOI: 10.1201/9781032702391-1

growth (Ince and Yetkin, 2023). For example, when the tip of a carrot is cut, it can regrow from the remaining root tissue. However, the cambium, a layer of actively dividing cells in stems and roots, contributes to secondary growth and wound healing (Partap et al., 2023).

6. *Seed Germination*: The natural process of plant regeneration involves the production of seeds. Seeds contain the genetic information needed to develop into new plants. When seeds germinate, they give rise to seedlings, ensuring the next generation of plants (Reed and Kent, 2022).
7. *Regeneration after Damage*: Some plants can regenerate from tissues damaged by herbivores or physical injury. For instance, certain grasses can regrow after being grazed by herbivores.

Understanding plant regeneration is vital for propagation in agriculture and ecological processes and the overall resilience of plant populations in natural environments (Vayabari and Zul, 2023).

The Pulianmackal et al. explores the widespread use of regeneration across various organisms, emphasizing its varying competence among species. While plant cells exhibit high regenerative potential due to developmental plasticity, maintaining stemness in mammalian cells is limited. Figure 1.1 shows how hormonal regulation and transcription factor expression influence de novo organ initiation in plants.

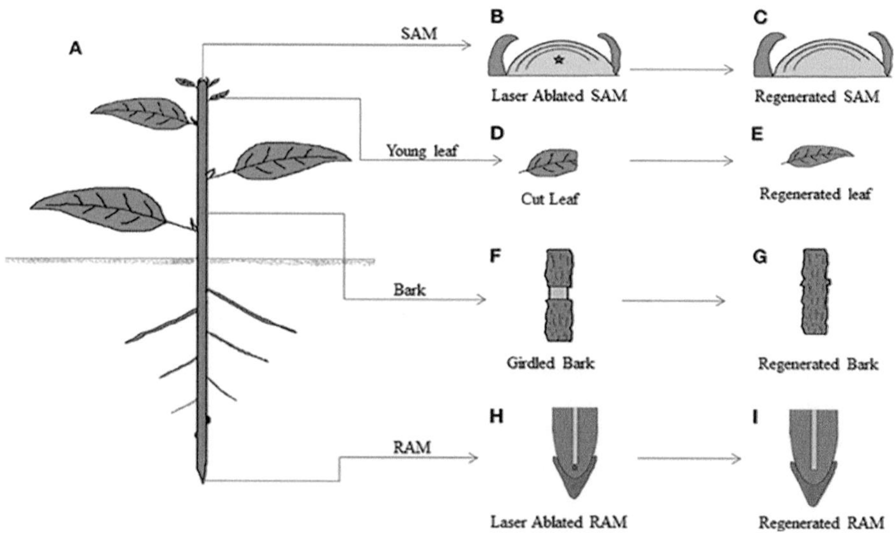

FIGURE 1.1 Illustration depicting the regenerative capabilities of different plant components. The plant exhibits significant regeneration: (a) Laser ablation of the shoot apical meristem (SAM) results in the regeneration of the wound without the need for external hormonal intervention (b, c). Likewise, various plant organs such as young leaves (d, e), bark (f, g), and root apical meristem (RAM) (h, i) demonstrate the ability to fully regenerate from an ablated organ without the use of additional hormones. (Image reprint from reference Pulianmackal and Abdul, 2014.)

1.1.2 Significance of Plant Regeneration

Plant regeneration is a fundamental biological process that allows plants to replace or repair damaged or lost tissues, ensuring their survival and adaptation. It encompasses various mechanisms through which plants can produce new organs, tissues, or whole plants, often in response to environmental cues, stress, or as part of their natural life cycle. Understanding plant regeneration is crucial for agricultural practices, ecological restoration, and the study of plant development.

1. *Survival and Adaptation*: Plant regeneration is essential for the survival of individual plants and the persistence of plant populations in changing environments. It enables plants to recover from injuries, herbivore damage, or adverse conditions, increasing their chances of survival (Lardon and Danny, 2020).
2. *Agricultural Propagation*: Farmers and horticulturists leverage the principles of plant regeneration for efficient and controlled propagation of crops. Techniques like vegetative propagation, grafting, and tissue culture replicate desirable plant traits and improve crop yields (Wang and Weihong, 2021).
3. *Biodiversity Conservation*: Understanding plant regeneration is crucial for biodiversity conservation. In ecosystems, the ability of plants to regenerate contributes to the overall health and diversity of plant communities, supporting various organisms and ecological processes (Furey and David, 2021).
4. *Ecological Restoration*: In ecological restoration projects, knowledge of plant regeneration is applied to rehabilitate degraded ecosystems. Restoration efforts often involve reintroducing native plant species and facilitating their regeneration to enhance ecosystem functionality (Carlucci and Pedro, 2020).
5. *Genetic Diversity and Evolution*: Plant regeneration plays a role in maintaining and generating genetic diversity within plant populations. This diversity is vital for evolutionary processes, enabling plants to adapt to changing environmental conditions and challenges over time (Rubio and Luis, 2020).
6. *Phytoremediation*: Certain plants can regenerate in contaminated environments. This property is exploited in phytoremediation, where plants are used to absorb, accumulate, or break down pollutants, contributing to environmental cleanup (Wei and Quyet, 2021).
7. *Seed Production and Germination*: Seeds and their subsequent germination are natural plant regeneration. Seeds contain the genetic material required for developing new plants, ensuring the continuation of plant species (Carrera-Castaño et al., 2020).
8. *Medicine and Biotechnology*: Plant regeneration techniques, particularly in vitro methods like tissue culture, are employed in biotechnology and the production of medicinal plants. This has applications in the pharmaceutical industry and the cultivation of plant-derived therapeutic compounds.

In summary, plant regeneration is a dynamic and essential process underpinning plant life's resilience, diversity, and sustainability. Its significance extends across various fields, impacting agriculture, ecology, biodiversity conservation, and advancements in biotechnology. Research in plant regeneration continues to uncover new insights, contributing to the effective management of plant resources and ecosystems (Srivastava and Mithilesh, 2020).

1.2 HISTORICAL PERSPECTIVE

Plant regeneration research has undergone significant milestones over the years, leading to a deeper understanding of the mechanisms and applications of this crucial biological process. Figure 1.2 illustrates essential regeneration systems in *Arabidopsis thaliana,* showcasing the reconstruction of shoot or root apical meristems following excision or laser ablation.

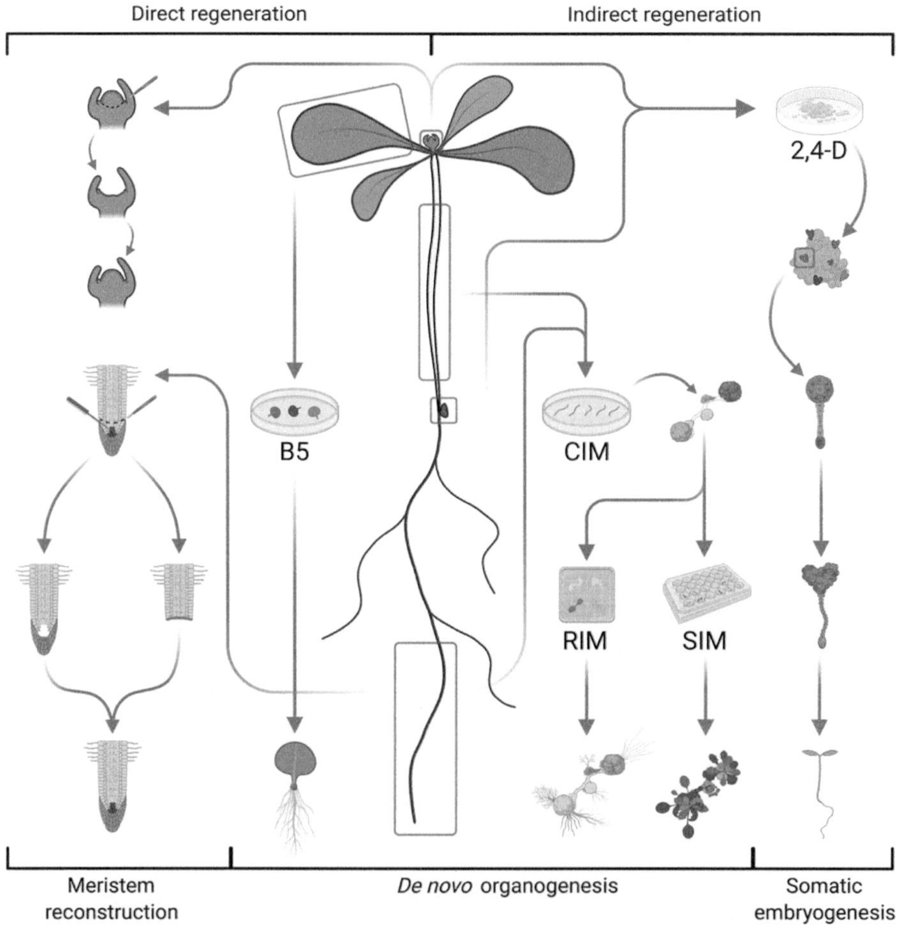

FIGURE 1.2 Essential regeneration systems in plant *Arabidopsis thaliana.* (Image reprint from reference Lardon and Danny, 2020.)

It also highlights adventitious rooting from excised leaves or cotyledons, *de novo* shoots or root organogenesis from root or hypocotyl explants, and somatic embryogenesis originating from immature zygotic embryos or shoot apices. Here are some critical milestones in plant regeneration research:

1. *Discovery of Totipotency (Late 19th Century)*: The concept of totipotency, introduced by German botanist Gottlieb Haberlandt in the late 19th century, laid the foundation for plant regeneration research. Totipotency refers to the ability of a single plant cell to give rise to an entire organism. The discovery that a single plant cell, under certain conditions, could regenerate into a whole plant opened avenues for tissue culture techniques and applications in agriculture (Kutschera and Ulrich, 2022).
2. *Introduction of Plant Tissue Culture (Early to Mid-20th Century)*: In the 20th century, researchers such as G. L. Stebbins and F.C. Steward pioneered plant tissue culture techniques, providing a controlled environment for the in vitro regeneration of plants from isolated tissues or cells. The development of tissue culture methods allowed for the mass production of genetically identical plants, contributing to advancements in agriculture, horticulture, and the propagation of ornamental plants (Sarma and Anwesha, 2023).
3. *Regeneration from Callus Cultures (1950s)*: F.C. Steward demonstrated the regeneration of whole plants from callus cultures—masses of undifferentiated cells. This finding highlighted the potential of manipulating plant cells to generate new tissues and organs. Callus cultures are widely used in tissue culture to propagate plants and produce transgenic organisms (Nurwahyuni and Isnaini, 2020).
4. *Discovery of Plant Growth Regulators (1950s–1960s)*: The identification and isolation of plant growth regulators, such as auxins and cytokinins, by scientists like Kenneth V. Thimann and Frederick C. Steward, provided insights into the hormonal control of plant regeneration. The balanced use of auxins and cytokinins in tissue culture media has become a standard practice for promoting cell division, organogenesis, and somatic embryogenesis in plant regeneration studies (Bhatla and Manju, 2018; 2023).
5. *Somatic Embryogenesis (Late 20th Century)*: The discovery of somatic embryogenesis, where embryos develop from somatic cells (non-reproductive cells), offered a new approach to plant regeneration and clonal propagation. Somatic embryogenesis produces genetically uniform plant material, such as in the mass propagation of coniferous tree species (Salaun et al., 2021; Yang and Xianlong, 2010).
6. *Genetic Transformation Techniques (1980s–1990s)*: The development of genetic transformation techniques, including Agrobacterium-mediated gene transfer and particle bombardment, allowed foreign genes into plant cells, paving the way for genetically modified (GM) plants. The creation of insect-resistant crops, like Bt cotton, through genetic engineering, is a notable application of plant regeneration research with implications for agriculture and pest control (Niedbała and Gniewko, 2021).

7. *Advancements in Seed Germination (20th Century Onward)*: Ongoing research has focused on understanding and improving seed germination processes, including the role of environmental factors and seed dormancy mechanisms. Studies on breaking seed dormancy in crops like sunflowers and certain fruits have practical implications for agriculture, ensuring optimal and uniform germination.
8. *Application in Conservation Biology (21st Century)*: Plant regeneration research has increasingly contributed to the conservation of endangered plant species. Techniques such as seed banking and in vitro propagation play a vital role in preserving genetic diversity. The successful regeneration of rare and endangered orchid species through tissue culture has aided their conservation and reintroduction into natural habitats.

These milestones collectively highlight the evolution of plant regeneration research, from fundamental discoveries of totipotency to practical applications in agriculture, biotechnology, and conservation. Ongoing advancements continue to expand our understanding of plant regeneration and its diverse applications in plant science and related fields (Yang and Liwen 2020; Sohindji and Deedi, 2020).

1.3 MECHANISMS OF PLANT REGENERATION

1.3.1 Cellular Processes Involved

Plant regeneration involves complex cellular processes that allow plants to replace damaged or lost tissues. Figure 1.3 exhibits various tissue culture techniques for plant regeneration, selection, and genetic transformation. Explants exposed to selection media can lead to direct organogenesis, as depicted in scenarios (a) and (b), or indirect organogenesis through an intermediate callus phase, as shown in (c) and (d). Moreover, callus derived from these processes can generate intact plantlets through an embryonic pathway. Alternatively, callus can be employed in suspension culture directly or through protoplast culture techniques, especially in genetic transformation efforts. This versatility in tissue culture methods provides diverse approaches for the regeneration and modification of plants in the context of genetic studies.

These processes encompass dedifferentiation, callus formation, cell division and proliferation, organogenesis, and re-differentiation. Here's a detailed explanation of the cellular processes involved in plant regeneration:

1. *Dedifferentiation*: Dedifferentiation is when specialized cells revert to a less specialized state or become undifferentiated. When a plant experiences damage or stress, specific cells at or near the injury site dedifferentiate (Wang and Yuman, 2022). This involves the loss of their specialized functions and the acquisition of a more primitive, undifferentiated state. Dedifferentiated cells are often referred to as meristematic cells (Partap et al., 2023). As Grafi et al. explore in Figure 1.4, the somatic cells, including parenchyma, collenchyma, or mesophyll cells, exhibit the capacity for reprogramming when exposed to internal or external signals. This reprogramming leads to dedifferentiation, where cells acquire a transient, stem cell-like state. This temporary state is characterized by global chromatin

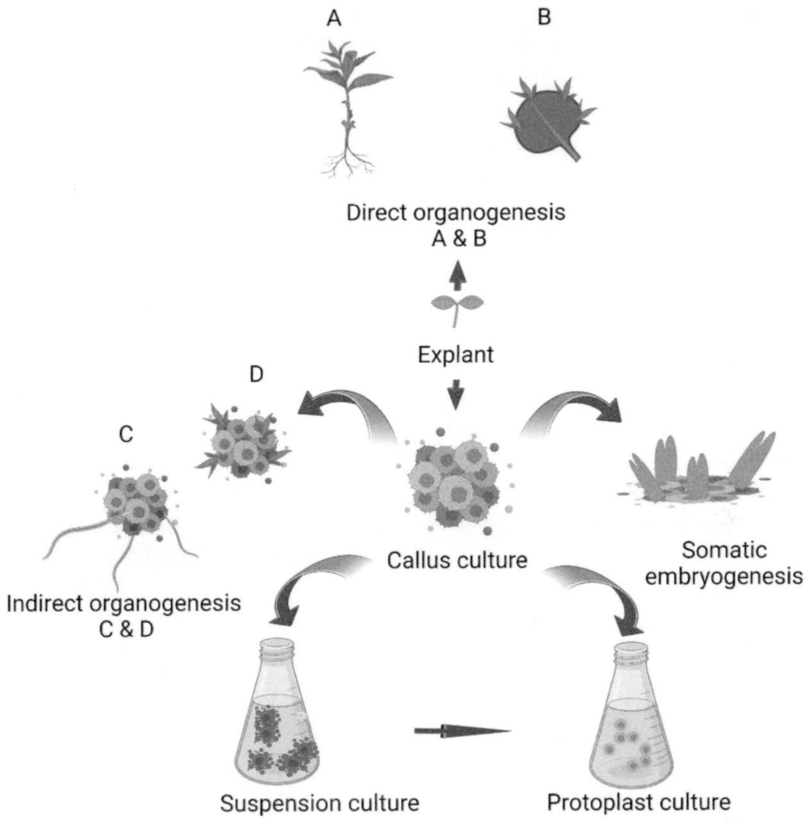

FIGURE 1.3 Different techniques for plant regeneration. (Image reprint from reference Wijerathna-Yapa and Akila, 2023.)

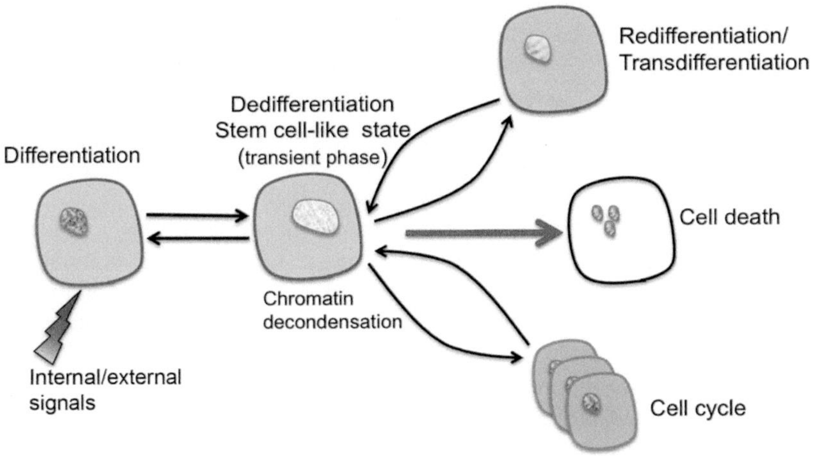

FIGURE 1.4 Cellular Dedifferentiation process in plant. (Image Reprint from reference Grafi and Assa, 2011.)

cell-like state. This temporary state is characterized by global chromatin decondensation, a key feature of stem cells. Depending on the specific stimulus, dedifferentiated cells can be directed to trans-differentiate or re-differentiate, re-enter the cell cycle, or undergo programmed cell death. The plasticity of somatic cells in response to various signals highlights their potential to undergo diverse fates, offering insights into cellular reprogramming and differentiation mechanisms.

2. *Callus Formation*: Callus is a mass of undifferentiated cells that forms at the site of injury or in tissue culture as a result of dedifferentiation. Dedifferentiated cells undergo rapid cell division, leading to the formation of a callus. The callus serves as a reservoir of pluripotent cells, capable of giving rise to various cell types (Shim et al., 2020). Callus is often induced deliberately in tissue culture to initiate regeneration (Suo et al., 2021).
3. *Cell Division and Proliferation*: Cell division is the process by which cells replicate, increasing their numbers. Within the callus, cells undergo active cell division to produce a large population of undifferentiated cells. This proliferation phase is crucial for generating sufficient cells that can differentiate into specific tissues and organs (Oh and Saxon, 2020).
4. *Organogenesis*: Organogenesis is the process through which specific organs, such as shoots, roots, or leaves, develop from undifferentiated cells (Hussain and Sajid, 2021). Some cells within the callus undergo organogenesis, differentiating into specific cell types that will form distinct plant organs. The hormonal balance, particularly the ratio of auxins to cytokinins, is pivotal in determining the kind of organ formed (Pernisová and Markéta, 2009).
5. *Embryogenesis*: Embryogenesis is the process by which an embryo develops from a fertilized egg or, in plants, from somatic cells. Plants can undergo somatic embryogenesis in specific regeneration pathways, especially in tissue culture. This involves the formation of structures resembling embryos from somatic cells. These structures, called somatic embryos, can potentially develop into complete plants (Salaun et al., 2021).
6. *Re-differentiation*: Re-differentiation is the process by which undifferentiated cells regain specialized functions and morph into specific cell types. In the final stages of plant regeneration, the cells that have undergone dedifferentiation and organogenesis re-differentiate into particular cell types. This process is essential for developing functional tissues and restoring normal plant structure and function (Gehrke and Paola, 2023).
7. *Role of Plant Hormones*: Plant hormones, including auxins, cytokinins, gibberellins, and abscisic acid, are crucial in regulating the cellular processes involved in plant regeneration. The balance and interaction of these hormones influence cell fate determination, cell division, and differentiation (Khan et al., 2023; Buttò and Annie, 2020).
8. *Environmental and External Factors*: Environmental factors, such as light, temperature, and nutrient availability, can influence the success of plant regeneration. External factors, including the application of growth regulators, tissue culture techniques, and physical injury, also play significant roles in modulating cellular responses during regeneration (Shimotohno and Shiori, 2021).

Understanding these cellular processes is essential for manipulating plant regeneration in applications such as tissue culture, genetic engineering, and the production of disease-free plant material. Researchers continually explore and refine these processes to enhance the efficiency of plant regeneration techniques for agricultural, horticultural, and conservation purposes (Zeng and Yiwen, 2020).

1.3.2 Factors Influencing Regeneration

1.3.2.1 Crop Genotype

Genotype plays a pivotal role in plant regeneration, significantly influencing the efficiency and success of the process. Different genotypes exhibit varying capacities for shoot regeneration, with some showing high responsiveness and others with much lower potential (Galán-Ávila and García-Fortea, 2024). The response to culture media, growth regulators, and explant types can be highly genotype-specific, necessitating tailored media optimization for each genotype. This variability highlights the importance of understanding genetic factors to improve regeneration outcomes, ensuring successful and efficient plant development across diverse genotypes (Abdullah and Elwira, 2021). Recent advances in genome-editing tools have shown significant potential for accelerating functional genomics and crop trait improvements. However, practical applications are limited by low efficiency and genotype dependence in plant transformation. Morphogenic transcription factors (MTFs) such as Baby boom, Wuschel2, growth-regulating factor5, growth-regulating factor4 and its cofactor grf-interacting factor1, and Wuschel-homeobox 5 related have been found to significantly enhance plant transformation efficiency and broaden the range of species and genotypes that can be transformed (Lee and Kan, 2023).

Tang et al. (2003) investigated the effects of culture media, explants, and genotypes on shoot regeneration in oilseed *Brassica* species. The highest shoot regeneration frequency was achieved using Murashige and Skoog medium with 3 mg/L 6-benzylaminopurine and 0.15 mg/L 1-naphthaleneacetic acid. Adding 2.5 mg/L of $AgNO_3$ significantly improved shoot regeneration in *B. napus*, with $Ag_2S_2O_3$ (10 mg/L) being even more effective than $AgNO_3$ (2.5 mg/L). Explant age, type, and carbon source also significantly influenced shoot regeneration, with 4-day-old cotyledonary explants showing the highest frequency and number of shoots per explant. Among the explants tested, cotyledons produced the highest shoot regeneration frequency (56.67%). Sucrose was the best carbon source, achieving a 76% regeneration frequency. Considerable variation was observed between and within Brassica species, with shoot regeneration frequencies ranging from 10% for cv. R5 (*B. rapa*) to 83.61% for cv. N1 (B. napus). Several cultivars *of B. napus, B. carinata*, and *B. juncea* showed frequencies higher than 70%. The number of shoots per explant varied significantly, especially in *B. rapa*. All induced shoots developed into normal, fertile plantlets identical to the source plants (Tang et al., 2003).

1.3.2.2 Environment

Environmental factors play a crucial role in influencing plant regeneration. These factors can affect the success of regeneration processes such as seed germination, tissue culture, and the natural regeneration of plants in their native habitats (Figure 1.5).

Here are some key environmental factors, along with examples of their influence on plant regeneration:

1. *Light*: Light is a critical environmental factor that influences plant regeneration, especially in terms of seed germination and the growth of seedlings. Light is necessary to germinate light-dependent seeds, such as those of many small-seeded herbs. Additionally, the quality and quantity of light can affect the morphology and development of regenerated plant tissues in tissue culture (Smale and Albert, 2020).
2. *Temperature*: Temperature significantly affects the rates of biochemical and physiological processes involved in plant regeneration. Different plant species have specific temperature requirements for optimal growth and regeneration. Some plant species require specific temperature ranges for successful seed germination. For instance, many temperate fruit tree seeds need a period of cold stratification to break dormancy and facilitate germination (Klupczyńska and Tomasz, 2021).
3. *Moisture and Water Availability*: Adequate moisture is crucial for various stages of plant regeneration, including seed germination, callus induction, and root development. Seeds often require a certain level of humidity to initiate germination (Walck et al., 2011). Maintaining appropriate moisture

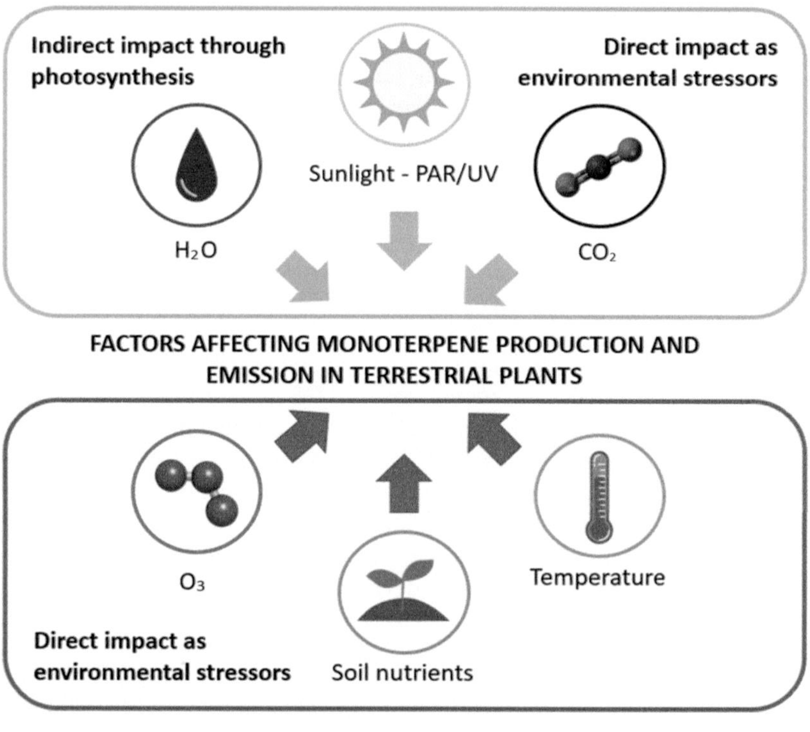

FIGURE 1.5 Effect of environmental factors on plant regeneration. (Image reprint from reference Malik et al., 2023.)

levels in the growth medium in tissue culture is essential for callus development and subsequent organogenesis (Calvia and Giacomo, 2023).

4. *Nutrient Availability*: Nutrients, including macronutrients and micronutrients, are essential for plant growth and regeneration. The availability of nutrients in the soil or growth medium affects cell division, differentiation, and overall plant development. In tissue culture, the nutrient medium's composition influences plant cells' regeneration potential. Adjusting the concentrations of specific nutrients can optimize the regeneration process (Dumani and Seyed, 2022; Cakmak and Patrick, 2023).
5. *Oxygen Levels*: Oxygen is crucial for cellular respiration and energy production in plant cells. Insufficient oxygen levels can hinder metabolic processes and affect regeneration. Waterlogged soils can lead to low oxygen availability, negatively impacting root development and regeneration. Adequate aeration is essential for successful plant regeneration in such conditions (Jethva and Romy, 2022).
6. *pH Levels*: Soil or growth medium pH can impact nutrient availability and influence enzymatic activities involved in plant regeneration processes. Some plants prefer acidic soils, while others thrive in alkaline conditions. Adjusting pH levels in tissue culture media is essential for optimizing the regeneration of specific plant species (Gillespie et al., 2020).
7. *Pathogens and Microbial Interactions*: Pathogens, fungi, or beneficial microorganisms can positively or negatively influence plant regeneration. Soil-borne pathogens can inhibit seed germination and affect the development of plant roots. Conversely, mycorrhizal associations may enhance nutrient uptake and promote the regeneration of certain plant species (Nishad and Resna, 2020).
8. *Biotic Interactions*: Interactions with other organisms, such as competition with neighboring plants or interactions with herbivores, can impact plant regeneration. In natural ecosystems, herbivore grazing may stimulate the regeneration of certain plant species by inducing the production of new shoots and leaves as a response to herbivory.

Understanding and managing these environmental factors are essential for successful plant regeneration in various contexts, including agriculture, horticulture, and ecosystem restoration. Researchers and practitioners consider these factors to optimize plant growth and regeneration conditions in specific plant species or systems (Waller et al., 2020).

1.4 APPLICATIONS IN AGRICULTURE AND HERBAL INDUSTRIES

1.4.1 Crop or Medicinal Plant Improvement through Regeneration

Plant regeneration techniques are extensively employed in agricultural and medicinal plant improvement programs. These techniques contribute to developing improved crop varieties with enhanced traits or the efficient production of medicinal plants with valuable secondary metabolites (Ozyigit and Ibrahim, 2023). One impactful application of plant regeneration in agriculture and herbal industries is enhancing the medicinal compound production of invaluable herbs. By leveraging plant regeneration techniques, mainly through tissue culture and genetic transformation, it is possible to optimize the biosynthesis of therapeutic compounds in medicinal plants. For instance, consider the cultivation of Ginseng (*Panax ginseng*), a renowned herb with various health-promoting properties.

The process begins by initiating tissue culture from selected explants, leading to the induction and proliferation of callus—a source of undifferentiated cells (Xu and Anjali, 2023). Through genetic transformation, specific genes associated with the biosynthesis of medicinal compounds are introduced into the callus, creating GM plants with an augmented capacity for compound production. High-yielding lines are selected by scaling up regeneration in controlled environments. The GM plants can be propagated for commercial cultivation upon successful field testing and agronomic evaluation. This application ensures a consistent and reliable yield of medicinal compounds and contributes to sustainable agriculture by maximizing the production of valuable bioactive constituents in medicinal herbs, meeting the demands of the herbal industries for high-quality and natural products (Akinbo and Silas, 2021).

However, incorporating GM crops presents significant challenges and considerations that warrant careful attention. Regulatory approval stands as a formidable hurdle, demanding a thorough evaluation of GM organisms' safety and environmental impact. Introducing such crops necessitates compliance with stringent regulatory frameworks, often involving extensive testing and assessments to ensure that the modified genes do not pose risks to ecosystems or human health (Bauer-Panskus and Juliana, 2020). Equally crucial is public acceptance, as societal attitudes and perceptions play a pivotal role in determining the success and adoption of GM crops.

Public awareness campaigns and transparent communication about the benefits and safety of these crops become essential in navigating public sentiments. Additionally, ensuring the stability of gene expression across multiple generations is paramount for the sustained success of GM crops. Genetic stability ensures that the desired traits persist over time, contributing to crops' long-term improvement in yield, resilience, and other agronomic characteristics. As the agricultural landscape evolves, addressing these challenges is vital for harnessing the full potential of GM crops to contribute to global food security and sustainable agriculture (Kumar and Geetika, 2020).

This application of plant regeneration in crop improvement illustrates how these techniques can be harnessed to enhance the resilience and productivity of crops, addressing challenges such as pest and disease management sustainably. Similar approaches can be adopted to improve medicinal plants to enhance the yield of bioactive compounds with therapeutic value (Feng and Frances, 2020).

1.5 ECOLOGICAL IMPACT

Plant regeneration plays a vital role in maintaining the structure, function, and resilience of ecosystems. It involves the processes by which plants recover and renew their populations, ensuring the sustainability of biodiversity and ecosystem services

(Day and Alison, 2020). Here's a detailed explanation with examples of the crucial role of plant regeneration in ecosystems:

1. *Biodiversity Maintenance*: Plant regeneration is fundamental to the maintenance of biodiversity within ecosystems. It ensures the continuous presence of various plant species, each contributing to the overall diversity and ecological balance. In a natural forest ecosystem, plant regeneration processes, such as seed germination and seedling establishment, support the coexistence of a diverse array of tree, shrub, and herb species (Capellesso and Anamaria, 2021).
2. *Ecological Succession*: Plant regeneration is a driving force behind ecological succession, the process through which an ecosystem transitions from one community structure to another over time. It allows for the replacement of pioneer species with more complex and stable communities. In a disturbed area, such as a recently abandoned agricultural field, plant regeneration starts with pioneer species like grasses and weeds. Over time, these are succeeded by shrubs and eventually by trees, leading to a climax community (Onésimo and Diego, 2021).
3. *Habitat Restoration*: Plant regeneration is crucial for the restoration of degraded habitats. It involves the reintroduction of native plant species, helping to rebuild ecosystems that human activities or natural disturbances have impacted. Wetland restoration projects often involve planting native vegetation or encouraging the regeneration of wetland plants to enhance habitat quality for various species and improve water quality (Genes and Luisa, 2022).
4. *Soil Stability and Erosion Control*: Plants contribute to soil stability through their root systems. Plant regeneration, particularly of deep-rooted species, helps prevent soil erosion, reduce the risk of landslides, and stabilize slopes. After a wildfire, the regeneration of vegetation, including grasses and shrubs, is crucial for stabilizing the soil and preventing erosion in the affected area (Soto and María, 2021).
5. *Carbon Sequestration*: Growing plants capture carbon dioxide through photosynthesis, contributing to carbon sequestration. Plant regeneration, especially in forests, supports the storage of carbon in plant biomass and soil organic matter. Forest regeneration after disturbances like logging or natural disasters facilitates the sequestration of carbon, helping mitigate the impacts of climate change (Elbasiouny and Hassan, 2022).
6. *Wildlife Habitat Provision*: Plant regeneration directly influences the availability of habitats for wildlife. Different plant communities provide varied habitats, supporting diverse animal species. In grassland ecosystems, the regeneration of native grass species is crucial for providing suitable habitat for herbivores like bison and deer, as well as their predators (Lacerte and Mathieu, 2021).
7. *Water Cycle Regulation*: Plants contribute to the regulation of the water cycle by absorbing and transpiring water. Plant regeneration helps maintain the balance of water resources in ecosystems. Riparian vegetation

along riverbanks plays a critical role in water cycle regulation by preventing erosion, filtering pollutants, and regulating water flow. The regeneration of riparian plants is vital for maintaining these ecosystem services (Allan et al., 2020; Douville and Krishnan, 2021).

8. *Resistance to Invasive Species*: Healthy, regenerating plant communities can resist the establishment and spread of invasive species, helping to maintain the integrity of native ecosystems. In a grassland ecosystem, the regeneration of native grass species may compete with and limit the spread of invasive plant species that could otherwise outcompete and displace native vegetation.

Understanding and supporting the processes of plant regeneration are essential for the conservation and sustainable management of ecosystems (Green and Edwin, 2021). Human activities that impact plant regeneration, such as deforestation, habitat destruction, or overgrazing, can have cascading effects on the overall health and resilience of ecosystems. Conservation efforts often involve strategies to enhance plant regeneration and ensure the long-term stability of natural habitats (Zhen and Bojie, 2014).

1.6 CHALLENGES AND FUTURE DIRECTIONS

1.6.1 Current Limitations in Plant Regeneration

1. *Species and Genotype Dependency*: The success of plant regeneration methods varies among different plant species and even within different genotypes of the same species. Some plants exhibit high regeneration potential, while others are more recalcitrant (Maren et al., 2022). While some varieties of maize (*Zea mays*) can be regenerated efficiently in tissue culture, other varieties may show low regeneration rates, complicating efforts for uniform regeneration across the entire species (Ishida and Yukoh, 2020).
2. *Recalcitrance of Woody Plants*: Woody plants, like many tree species, often exhibit recalcitrance to in vitro regeneration. The regeneration of shoots and roots from woody explants is often challenging. Coniferous trees like pine and spruce may be recalcitrant to regeneration, posing challenges in forest tree improvement programs where the efficient regeneration of these species is crucial (Lv and Fei, 2021).
3. *Somatic Mutation Accumulation*: During in vitro regeneration, somatic mutations can accumulate, leading to genetic instability. This can result in unpredictable changes in plant traits and characteristics. Regenerated plants of some crops, like potato (*Solanum tuberosum*), may exhibit variations in tuber size, shape, or color due to somaclonal variations introduced during regeneration (Aksoy and Emre, 2021).
4. *Epigenetic Changes*: In vitro regeneration processes can induce epigenetic modifications, potentially altering gene expression patterns and phenotypic traits. In rice (*Oryza sativa*) regeneration, changes in DNA methylation patterns have been observed, influencing the expression of genes related to plant development and stress responses (Wang et al., 2023).

5. *Low Regeneration Efficiency*: The efficiency of plant regeneration methods can be low, resulting in a limited number of successfully regenerated plants. Regeneration efficiency in crops like wheat (*Triticum aestivum*) varies between cultivars, and optimizing regeneration protocols is essential for obtaining higher numbers of healthy regenerants (Yadav and Kapil, 2020).
6. *Environmental Sensitivity*: Environmental factors, such as temperature, light, and humidity, significantly influence plant regeneration success. Suboptimal conditions may lead to low regeneration rates or abnormal development (Campbell and Jackie, 2020). Regeneration of certain ornamental plants, like orchids, is highly influenced by the specific environmental conditions in tissue culture, making standardization challenging (Mose and Budi, 2020).
7. *Lack of Standardization*: There is a lack of standardized regeneration protocols for many plant species, leading to variations in techniques, growth media, and hormonal formulations used across different laboratories (Sahoo and Amit, 2011). The regeneration of *Arabidopsis thaliana*, a model plant species, is well-established with standardized protocols. However, for non-model species, there is often a lack of consensus on the optimal regeneration procedures (Karesch and Roland, 1991).
8. *Regeneration of Specific Tissues*: Certain tissues, such as flowers or specialized organs, are more challenging to regenerate than others. Regenerating flowers in crops like soybean (*Glycine max*) is particularly challenging, limiting the exploration of floral traits and seed development applications.

Addressing these limitations involves ongoing research to understand the molecular and physiological mechanisms behind plant regeneration, develop more robust regeneration protocols, and enhance the applicability of these methods across a wider range of plant species and genotypes. Overcoming these challenges is crucial for the successful integration of plant regeneration techniques in various fields, including agriculture, forestry, and conservation (Xu and Lixiao, 2020).

1.6.2 Emerging Technologies and Innovation

Advancements in biotechnology and plant science have led to the emergence of innovative technologies in plant regeneration. These technologies aim to overcome existing limitations, enhance efficiency, and broaden the applicability of plant regeneration methods (Sijacic-Nikolic et al., 2020). Here are some notable examples of emerging technologies and innovations:

1. *CRISPR-Cas9 Genome Editing*: CRISPR-Cas9 allows precise and targeted modification of plant genomes, enabling the introduction or deletion of specific genes associated with regeneration and developmental processes. CRISPR-Cas9 has been used to enhance the regeneration potential of crops like wheat by modifying key genes involved in tissue differentiation and organ development (Kim and Burcu, 2018).

2. *Synthetic Biology for Plant Regeneration*: Synthetic biology involves designing and constructing new biological entities or redesigning existing biological systems. In plant regeneration, synthetic biology approaches aim to engineer plants with improved regeneration capabilities. Researchers are designing synthetic gene circuits that can enhance the efficiency of regeneration pathways, leading to more predictable and controllable outcomes (Goold and Philip, 2018).
3. *Advanced Tissue Culture Techniques*: Innovations in tissue culture methods, such as temporary immersion systems, bioreactors, and micro-propagation, improve the efficiency and scalability of plant regeneration. Temporary immersion systems, which involve the periodic immersion of explants in liquid culture media, have been successfully applied to enhance shoot and root regeneration in various plant species (Brar and Khush, 2021).
4. *Use of Nanomaterials in Plant Regeneration*: Nanomaterials, including nanoparticles and nanotubes, are being explored for their potential to improve plant regeneration by influencing growth and development at the cellular and molecular levels. The application of nanomaterials in tissue culture media has been shown to enhance the regeneration efficiency of certain crops, promoting better shoot and root development (Saleem and Haleema, 2020).
5. *Optogenetics for Plant Manipulation*: Optogenetics involves the use of light-sensitive proteins to control cellular processes with high spatial and temporal precision. In plant regeneration, optogenetic tools are employed to modulate gene expression and developmental pathways. Optogenetic manipulation of key regulatory genes in plants has been used to control the timing and extent of cell differentiation during regeneration processes (Christie and Matias, 2021).
6. *Deciphering Signaling Pathways*: Advances in omics technologies, such as genomics, transcriptomics, and proteomics, allow researchers to decipher complex signaling pathways involved in plant regeneration. Comprehensive analysis of gene expression profiles during regeneration in model plants like Arabidopsis helps identify key regulatory genes and pathways, providing insights for targeted manipulation in other plant species (Dai and Shen, 2022).
7. *Use of Plant Growth-Promoting Microorganisms (PGPMs)*: Harnessing the potential of beneficial microorganisms to enhance plant growth and development, including regeneration. Inoculating plant tissues with specific strains of mycorrhizal fungi or plant growth-promoting bacteria has been shown to positively influence the regeneration capacity of plants by improving nutrient uptake and overall plant health (Hakim and Tahir, 2021).
8. *High-Throughput Screening for Regeneration Traits*: High-throughput screening methods enable the rapid and parallel assessment of large plant samples for regeneration traits. Automated systems that can screen plant populations for traits related to regeneration efficiency, such as shoot initiation rates or root development, allow researchers to identify promising genotypes for further study.

These emerging technologies hold great promise for advancing our understanding of plant regeneration processes and enhancing the efficiency of regeneration methods. As these innovations mature, they will contribute significantly to crop improvement, conservation efforts, and the sustainable production of plant-based resources (Hou and Yingji, 2020).

1.7 CONCLUSION

In conclusion, plant regeneration delves into the intricate processes by which plants renew and restore themselves, emphasizing their profound significance in ecological, agricultural, and biotechnological contexts. Plant regeneration encompasses a series of cellular events, including dedifferentiation, callus formation, organogenesis, and re-differentiation, orchestrated by complex molecular and hormonal networks. The significance of plant regeneration is multifaceted. In natural ecosystems, it plays a pivotal role in maintaining biodiversity, supporting ecological succession, and restoring degraded habitats. In agriculture, plant regeneration is a linchpin for crop improvement, enabling the development of varieties with enhanced traits such as disease resistance, stress tolerance, and improved yield. Moreover, plant regeneration is a cornerstone in biotechnological applications, facilitating tissue culture, genetic engineering, and the propagation of elite plant lines.

The exploration of milestones and notable discoveries in plant regeneration research highlights the evolving understanding of these processes and their potential applications. From the early observations of totipotency to contemporary advancements in genetic editing technologies, researchers have continually pushed the boundaries of our knowledge, unlocking new possibilities for manipulating plant regeneration for practical purposes. However, this chapter also underscores the current limitations in plant regeneration, ranging from genotype dependency to environmental sensitivity. Acknowledging these challenges is crucial for steering future research toward addressing these constraints and refining existing techniques. The integration of emerging technologies, such as CRISPR-Cas9 genome editing, synthetic biology, and advanced tissue culture methods, promises to overcome these limitations and usher in a new era of precision and efficiency in plant regeneration practices. As we navigate the dynamic landscape of plant regeneration research, it becomes evident that this field holds immense potential for addressing global challenges, including food security, environmental conservation, and sustainable resource utilization.

The continued exploration of plant regeneration not only enriches our understanding of fundamental biological processes but also empowers us to harness the resilience and adaptability of plants for the benefit of ecosystems and human societies alike. Through the integration of innovative technologies and a deeper comprehension of the intricacies involved, the future of plant regeneration holds the promise of unlocking new possibilities for sustainable agriculture, biodiversity conservation, and biotechnological innovation.

REFERENCES

Abdullah, Mohib, Elwira Sliwinska, Grzegorz Góralski, Piotr Latocha, Monika Tuleja, Patrycja Widyna, and Marzena Popielarska-Konieczna. "Effect of medium composition, genotype and age of explant on the regeneration of hexaploid plants from endosperm culture of tetraploid kiwiberry (*Actinidia arguta*)." *Plant Cell, Tissue and Organ Culture (PCTOC)* 147, 3 (2021): 569–582.

Akinbo, Olalekan, Silas Obukosia, Jeremy Ouedraogo, Woldeyesus Sinebo, Moussa Savadogo, Samuel Timpo, Ruth Mbabazi, Karim Maredia, Diran Makinde, and Aggrey Ambali. "Commercial release of genetically modified crops in Africa: interface between biosafety regulatory systems and varietal release systems." *Frontiers in Plant Science* 12 (2021): 605937.

Aksoy, Emre, Ufuk Demirel, Allah Bakhsh, Muhammad Abu Bakar Zia, Muhammad Naeem, Faisal Saeed, Sevgi Çalışkan, and Mehmet Emin Çalışkan. "Recent advances in potato (*Solanum tuberosum* L.) breeding." In *Advances in Plant Breeding Strategies: Vegetable Crops: Volume 8: Bulbs, Roots and Tubers*, pp. 409–487, 2021.

Allan, Richard P., Mathew Barlow, Michael P. Byrne, Annalisa Cherchi, Hervé Douville, Hayley J. Fowler, Thian Y. Gan et al. "Advances in understanding large-scale responses of the water cycle to climate change." *Annals of the New York Academy of Sciences* 1472, 1 (2020): 49–75.

Anjarwalla, Parveen, Daniel Ofori, Alfred Owino, David Matuku, Winnie Adika, Kennly Njogu, and Katja Kehlenbeck. "Testing different grafting methods for vegetative propagation of baobab (*Adansonia digitata* L.) in Kenya to assist its domestication and promote cultivation." *Forests, Trees and Livelihoods* 26, 2 (2017): 85–95.

Bansal, Sangita, Manoj K. Sharma, Parampara Joshi, Era V. Malhotra, M. Latha, and S. K. Malik. "An efficient direct organogenesis protocol for in vitro clonal propagation of *Rubia cordifolia* L." *Industrial Crops and Products* 208 (2024): 117856.

Bauer-Panskus, Andreas, Juliana Miyazaki, Katharina Kawall, and Christoph Then. "Risk assessment of genetically engineered plants that can persist and propagate in the environment." *Environmental Sciences Europe* 32, 1 (2020): 1–15.

Bhatla, Satish C., and Manju A. Lal. "Plant growth regulators: an overview." In *Plant Physiology, Development and Metabolism*, pp. 391–398, 2023.

Bhatla, Satish C., Manju A. Lal, and Satish C. Bhatla. "Recently discovered plant growth regulators." In *Plant Physiology, Development and Metabolism*, pp. 681–728, 2018.

Bidabadi, Siamak Shirani, and S. Mohan Jain. "Cellular, molecular, and physiological aspects of in vitro plant regeneration." *Plants* 9, 6 (2020): 702.

Brar, D. S., and Gurdev S. Khush. "Cell and tissue culture for plant improvement." In *Mechanisms of Plant Growth and Improved Productivity Modern Approaches*, pp. 229–278. CRC Press, 2021.

Bredmose, Niels, and Jürgen Hansen. "Regeneration, growth and flowering of cut rose cultivars as affected by propagation material and method." *Scientia Horticulturae* 64, 1–2 (1995): 103–111.

Buttò, Valentina, Annie Deslauriers, Sergio Rossi, Philippe Rozenberg, Vladimir Shishov, and Hubert Morin. "The role of plant hormones in tree-ring formation." *Trees* 34 (2020): 315–335.

Cakmak, Ismail, Patrick Brown, José M. Colmenero-Flores, Søren Husted, Bahar Y. Kutman, Miroslav Nikolic, Zed Rengel, Sidsel B. Schmidt, and Fang-Jie Zhao. "Micronutrients." In *Marschner's Mineral Nutrition of Plants*, pp. 283–385. Academic Press, 2023.

Calvia, Giacomo, Paolo Casula, Emmanuele Farris, Giuseppe Fenu, Sergio Fantini, and Gianluigi Bacchetta. "Shrub cover and soil moisture affect *Taxus baccata* L. regeneration at its southern range." *Plants* 12, 9 (2023): 1819.

Campbell, Dana L. M., Jackie Ouzman, Damian Mowat, Jim M. Lea, Caroline Lee, and Rick S. Llewellyn. "Virtual fencing technology excludes beef cattle from an environmentally sensitive area." *Animals* 10, 6 (2020): 1069.

Capellesso, Elivane Salete, Anamaria Cequinel, Renato Marques, Tanise Luisa Sausen, Cimélio Bayer, and Marcia Cristina Mendes Marques. "Co-benefits in biodiversity conservation and carbon stock during forest regeneration in a preserved tropical landscape." *Forest Ecology and Management* 492 (2021): 119222.

Carlucci, Marcos B., Pedro H. S. Brancalion, Ricardo R. Rodrigues, Rafael Loyola, and Marcus V. Cianciaruso. "Functional traits and ecosystem services in ecological restoration." *Restoration Ecology* 28, 6 (2020): 1372–1383.

Carrera-Castaño, Gerardo, Julián Calleja-Cabrera, Mónica Pernas, Luis Gómez, and Luis Oñate-Sánchez. "An updated overview on the regulation of seed germination." *Plants* 9, 6 (2020): 703.

Christie, John M., and Matias D. Zurbriggen. "Optogenetics in plants." *New Phytologist* 229, 6 (2021): 3108–3115.

Dai, Xiaofeng, and Li Shen. "Advances and trends in omics technology development." *Frontiers in Medicine* 9 (2022): 911861.

Day, Nicola J., Alison L. White, Jill F. Johnstone, Geneviève É. Degré-Timmons, Steven G. Cumming, Michelle C. Mack, Merritt R. Turetsky, Xanthe J. Walker, and Jennifer L. Baltzer. "Fire characteristics and environmental conditions shape plant communities via regeneration strategy." *Ecography* 43, 10 (2020): 1464–1474.

Douville, H., K. Raghavan, J. Renwick, R.P. Allan, P.A. Arias, M. Barlow, R. Cerezo-Mota, A. Cherchi, T.Y. Gan, J. Gergis, D. Jiang, A. Khan, W. Pokam Mba, D. Rosenfeld, J. Tierney, and O. Zolina, "2021: Water cycle changes." In *Climate Change 2021: The Physical Science Basis. Contribution of Working Group I to the Sixth Assessment Report of the Intergovernmental Panel on Climate Change*, Masson-Delmotte, V., P. Zhai, A. Pirani, S. L. Connors, C. Pean, S. Berger, N. Caud, Y. Chen, L. Goldfarb, M. I. Gomis, M. Huang, K. Leitzell, E. Lonnoy, J. B. R. Matthews, T. K. Maycock, T. Waterfield, O. Yelekci, R. Yu, and B. Zhou (eds.), pp. 1055–1210. Cambridge, United Kingdom and New York, NY: Cambridge University Press, 2021., doi:10.1017/9781009157896.010.

Dumani, Yasin, Seyed Mohammad Mahdi Mortazavian, Ali Izadi-Darbandi, and Hossein Ramshini. "Effect of macronutrients and micronutrients on biochemical properties in Paulownia shantung." *Plant Cell, Tissue and Organ Culture (PCTOC)* 151, 3 (2022): 483–495.

Elbasiouny, Heba, Hassan El-Ramady, Fathy Elbehiry, Vishnu D. Rajput, Tatiana Minkina, and Saglara Mandzhieva. "Plant nutrition under climate change and soil carbon sequestration." *Sustainability* 14, 2 (2022): 914.

Feng, Guoping, Frances E. Jensen, Henry T. Greely, Hideyuki Okano, Stefan Treue, Angela C. Roberts, James G. Fox et al. "Opportunities and limitations of genetically modified nonhuman primate models for neuroscience research." *Proceedings of the National Academy of Sciences of the United States of America* 117, 39 (2020): 24022–24031.

Galán-Ávila, Alberto, Edgar García-Fortea, Jaime Prohens, and Francisco Javier Herraiz. "Development of a direct in vitro plant regeneration protocol from *Cannabis sativa* L. seedling explants: developmental morphology of shoot regeneration and ploidy level of regenerated plants." *Frontiers in Plant Science.* Accessed July 23, 2024. https://pubmed.ncbi.nlm.nih.gov/32670304.

Furey, George N., and David Tilman. "Plant biodiversity and the regeneration of soil fertility." *Proceedings of the National Academy of Sciences of the United States of America* 118, 49 (2021): e2111321118.

Gehrke, Fabienne, Paola Ruiz-Duarte, Angelina Schindele, Sebastian Wolf, and Holger Puchta. "An inducible CRISPR-Kill system for temporally controlled cell type-specific cell ablation in Arabidopsis thaliana." *New Phytologist* 239, 5 (2023): 2041–2052.

Genes, Luisa, and Rodolfo Dirzo. "Restoration of plant-animal interactions in terrestrial ecosystems." *Biological Conservation* 265 (2022): 109393.

Gillespie, Daniel P., Chieri Kubota, and Sally A. Miller. "Effects of low pH of hydroponic nutrient solution on plant growth, nutrient uptake, and root rot disease incidence of basil (*Ocimum basilicum* L.)." *HortScience* 55, 8 (2020): 1251–1258.

Goold, Hugh Douglas, Philip Wright, and Deborah Hailstones. "Emerging opportunities for synthetic biology in agriculture." *Genes* 9, 7 (2018): 341.

Grafi, Gideon, Assa Florentin, Vanessa Ransbotyn, and Yakov Morgenstern. "The stem cell state in plant development and in response to stress." *Frontiers in Plant Science* 2 (2011): 53.

Green, Stephanie J., and Edwin D. Grosholz. "Functional eradication as a framework for invasive species control." *Frontiers in Ecology and the Environment* 19, 2 (2021): 98–107.

Guo, Lei, Madison Plunkert, Xi Luo, and Zhongchi Liu. "Developmental regulation of stolon and rhizome." *Current Opinion in Plant Biology* 59 (2021): 101970.

Hakim, Sughra, Tahir Naqqash, Muhammad Shoib Nawaz, Iqra Laraib, Muhammad Jawad Siddique, Rabisa Zia, Muhammad Sajjad Mirza, and Asma Imran. "Rhizosphere engineering with plant growth-promoting microorganisms for agriculture and ecological sustainability." *Frontiers in Sustainable Food Systems* 5 (2021): 617157.

Hou, Jinyan, Yingji Mao, Pengfei Su, Dacheng Wang, Xue Chen, Shengwei Huang, Jun Ni, Weiwei Zhao, and Lifang Wu. "A high throughput plant regeneration system from shoot stems of *Sapium sebiferum* Roxb., a potential multipurpose bioenergy tree." *Industrial Crops and Products* 154 (2020): 112653.

Hussain, Sajid, Satyabrata Nanda, Junhua Zhang, Muhammad Ishaq Asif Rehmani, Muhammad Suleman, Gaojie Li, and Hongwei Hou. "Auxin and cytokinin interplay during leaf morphogenesis and phyllotaxy." *Plants* 10, 8 (2021): 1732.

Ince, Yetkin Çaka, and Keiko Sugimoto. "Illuminating the path to shoot meristem regeneration: molecular insights into reprogramming cells into stem cells." *Current Opinion in Plant Biology* (2023): 102452.

Ishida, Yuji, Yukoh Hiei, and Toshihiko Komari. "Tissue culture protocols for gene transfer and editing in maize (*Zea mays* L.)." *Plant Biotechnology* 37, 2 (2020): 121–128.

Jethva, Jay, Romy R. Schmidt, Margret Sauter, and Jennifer Selinski. "Try or die: dynamics of plant respiration and how to survive low oxygen conditions." *Plants* 11, 2 (2022): 205.

Karesch, Hans, Roland Bilang, and Ingo Potrykus. "*Arabidopsis thaliana*: protocol for plant regeneration from protoplasts." *Plant Cell Reports* 9 (1991): 575–578.

Khan, M. Iqbal R., Sarika Kumari, Faroza Nazir, Risheek Rahul Khanna, Ravi Gupta, and Himanshu Chhillar. "Defensive role of plant hormones in advancing abiotic stress-resistant rice plants." *Rice Science* 30, 1 (2023): 15–35.

Kim, Dongjin, Burcu Alptekin, and Hikmet Budak. "CRISPR/Cas9 genome editing in wheat." *Functional & Integrative Genomics* 18 (2018): 31–41.

Klupczyńska, Ewelina A., and Tomasz A. Pawłowski. "Regulation of seed dormancy and germination mechanisms in a changing environment." *International Journal of Molecular Sciences* 22, 3 (2021): 1357.

Kumar, Krishan, Geetika Gambhir, Abhishek Dass, Amit Kumar Tripathi, Alla Singh, Abhishek Kumar Jha, Pranjal Yadava, Mukesh Choudhary, and Sujay Rakshit. "Genetically modified crops: current status and future prospects." *Planta* 251 (2020): 1–27.

Kutschera, Ulrich, and Peter M. Ray. "Forever young: stem cell and plant regeneration one century after Haberlandt 1921." *Protoplasma* 259, 1 (2022): 3–18.

Lacerte, Rebecca, Mathieu Leblond, and Martin-Hugues St-Laurent. "Determinants of vegetation regeneration on forest roads following restoration treatments: implications for boreal caribou conservation." *Restoration Ecology* 29, 7 (2021): e13414.

Lardon, Robin, and Danny Geelen. "Natural variation in plant pluripotency and regeneration." *Plants* 9, 10 (2020): 1261.

Lee, Keunsub, and Kan Wang. "Strategies for genotype-flexible plant transformation." *Current Opinion in Biotechnology* 79 (2023): 102848.

Lv, Zhengyi, Fei Liu, Youbing Zhang, Yuanyuan Tu, Peng Chen, and Liangcai Peng. "Ecologically adaptable *Populus simonii* is specific for recalcitrance-reduced lignocellulose and largely enhanced enzymatic saccharification among woody plants." *Gcb Bioenergy* 13, 2 (2021): 348–360.

Malik, Tanzil Gaffar, Lokesh Kumar Sahu, Mansi Gupta, Bilal Ahmad Mir, Triratnesh Gajbhiye, Rashmi Dubey, Andrea Clavijo McCormick, and Sudhir Kumar Pandey. "Environmental factors affecting monoterpene emissions from terrestrial vegetation." *Plants* 12, 17 (2023): 3146.

Maren, Nathan A., Hui Duan, Kedong Da, G. Craig Yencho, Thomas G. Ranney, and Wusheng Liu. "Genotype-independent plant transformation." *Horticulture Research* 9 (2022): uhac047.

Mose, Windi, Budi Setiadi Daryono, Ari Indrianto, Aziz Purwantoro, and Endang Semiarti. "Direct somatic embryogenesis and regeneration of an Indonesian orchid *Phalaenopsis amabilis* (L.) blume under a variety of plant growth regulators, light regime, and organic substances." *Jordan Journal of Biological Sciences* 13, 4 (2020).

Niedbała, Gniewko, Mohsen Niazian, and Paolo Sabbatini. "Modeling agrobacterium-mediated gene transformation of tobacco (*Nicotiana tabacum*)—a model plant for gene transformation studies." *Frontiers in Plant Science* 12 (2021): 695110.

Nishad, Resna, Talaat Ahmed, Vattakandy Jasin Rahman, and Abdul Kareem. "Modulation of plant defense system in response to microbial interactions." *Frontiers in Microbiology* 11 (2020): 1298.

Nurwahyuni, Isnaini, Manihar Situmorang, and Riyanto Sinaga. "Plant regeneration through callus cultures from leaf explant of Sumatra Benzoin (*Styrax benzoin*)." *International Journal of Forestry Research* 2020 (2020): 1–7.

Oh, Man-Ho, Saxon H. Honey, and Frans E. Tax. "The control of cell expansion, cell division, and vascular development by brassinosteroids: a historical perspective." *International Journal of Molecular Sciences* 21, 5 (2020): 1743.

Onesimo, Cecilia M. G., Diego D. Dias, Marina Vale Beirao, Alessandra R. Kozovits, and Maria Cristina T. B. Messias. "Ecological succession in areas degraded by bauxite mining indicates successful use of topsoil." *Restoration Ecology* 29, 1 (2021): e13303.

Ozyigit, Ibrahim Ilker, Ilhan Dogan, Asli Hocaoglu-Ozyigit, Bestenur Yalcin, Aysegul Erdogan, Ibrahim Ertugrul Yalcin, Evren Cabi, and Yilmaz Kaya. "Production of secondary metabolites using tissue culture-based biotechnological applications." *Frontiers in Plant Science* 14 (2023): 1132555.

Partap, Mahinder, Ashish R. Warghat, and Sanjay Kumar. "Cambial meristematic cell culture: a sustainable technology toward in vitro specialized metabolites production." *Critical Reviews in Biotechnology* 43, 5 (2023): 734–752.

Pernisová, Markéta, Petr Klíma, Jakub Horák, Martina Válková, Jiří Malbeck, Přemysl Souček, Pavel Reichman et al. "Cytokinins modulate auxin-induced organogenesis in plants via regulation of the auxin efflux." *Proceedings of the National Academy of Sciences of the United States of America* 106, 9 (2009): 3609–3614.

Pulianmackal, Ajai J., Abdul V. K. Kareem, Kavya Durgaprasad, Zankhana B. Trivedi, and Kalika Prasad. "Competence and regulatory interactions during regeneration in plants." *Frontiers in Plant Science* 5 (2014): 142.

Reed, Reagan C., Kent J. Bradford, and Imtiyaz Khanday. "Seed germination and vigor: ensuring crop sustainability in a changing climate." *Heredity* 128, 6 (2022): 450–459.

Rubio, Luis, Luis Galipienso, and Inmaculada Ferriol. "Detection of plant viruses and disease management: relevance of genetic diversity and evolution." *Frontiers in Plant Science* 11 (2020): 1092.

Sahoo, Khirod K., Amit K. Tripathi, Ashwani Pareek, Sudhir K. Sopory, and Sneh L. Singla-Pareek. "An improved protocol for efficient transformation and regeneration of diverse indica rice cultivars." *Plant Methods* 7, 1 (2011): 1–11.

Salaun, Camille, Loic Lepiniec, and Bertrand Dubreucq. "Genetic and molecular control of somatic embryogenesis." *Plants* 10, 7 (2021): 1467.

Saleem, Haleema, and Syed Javaid Zaidi. "Recent developments in the application of nanomaterials in agroecosystems." *Nanomaterials* 10, 12 (2020): 2411.

Sarma, M. K., Anwesha Ananya Sharma, Kajal Samantara, and Shabir H. Wani. "In vitro techniques in plant breeding." In *Advanced Crop Improvement, Volume 1: Theory and Practice*, pp. 185–215. Cham: Springer International Publishing, 2023.

Shim, Sangrea, Hee Kyoung Kim, Soon Hyung Bae, Hoonyoung Lee, Hyo Ju Lee, Yu Jin Jung, and Pil Joon Seo. "Transcriptome comparison between pluripotent and non-pluripotent calli derived from mature rice seeds." *Scientific Reports* 10, 1 (2020): 21257.

Shimotohno, Akie, Shiori S. Aki, Naoki Takahashi, and Masaaki Umeda. "Regulation of the plant cell cycle in response to hormones and the environment." *Annual Review of Plant Biology* 72 (2021): 273–296.

Sijacic-Nikolic, M., A. Ivanova, and N. Sirotkina. "Economic advantages of innovative technologies for accelerated forest regeneration in Russia." In *IOP Conference Series: Earth and Environmental Science*, vol. 595, no. 1, p. 012046. IOP Publishing, 2020.

Smale, Dan A., Albert Pessarrodona, Nathan King, Michael T. Burrows, Anna Yunnie, Thomas Vance, and Pippa Moore. "Environmental factors influencing primary productivity of the forest-forming kelp *Laminaria hyperborea* in the northeast Atlantic." *Scientific Reports* 10, 1 (2020): 12161.

Sohindji, Fernand S., Deedi E. O. Sogbohossou, Herbaud P. F. Zohoungbogbo, Carlos A. Houdegbe, and Enoch G. Achigan-Dako. "Understanding molecular mechanisms of seed dormancy for improved germination in traditional leafy vegetables: an overview." *Agronomy* 10, 1 (2020): 57.

Soto, Raquel Luján, María Martínez-Mena, Mamen Cuéllar Padilla, and Joris de Vente. "Restoring soil quality of woody agroecosystems in Mediterranean drylands through regenerative agriculture." *Agriculture, Ecosystems & Environment* 306 (2021): 107191.

Srivastava, Priyanka, Mithilesh Singh, and Rakhi Chaturvedi. "Herbal medicine and biotechnology for the benefit of human health." In *Animal Biotechnology*, pp. 613–629. Academic Press, 2020.

Suo, Jingqi, Chenlu Zhou, Zhanghui Zeng, Xipu Li, Hongwu Bian, Junhui Wang, Muyuan Zhu, and Ning Han. "Identification of regulatory factors promoting embryogenic callus formation in barley through transcriptome analysis." *BMC Plant Biology* 21 (2021): 1–19.

Tang, G. X., W. J. Zhou, H. Z. Li, B. Z. Mao, Z. H. He, and K. Yoneyama. "Medium, explant and genotype factors influencing shoot regeneration in oilseed *Brassica* spp." *Journal of Agronomy and Crop Science* 189, 5 (2003): 351–358.

Vayabari, Diwiyaa A. G., Zul Ilham, Norsharina Md Saad, Siti Rokhiyah Ahmad Usuldin, Danial'Aizat Norhisham, Muhamad Hafiz Abd Rahim, and Wan Abd Al Qadr Imad Wan-Mohtar. "Cultivation strategies of kenaf (*Hibiscus cannabinus* L.) as a future approach in Malaysian agriculture industry." *Horticulturae* 9, 8 (2023): 925.

Walck, Jeffrey L., Siti N. Hidayati, Kingsley W. Dixon, K. E. N. Thompson, and Peter Poschlod. "Climate change and plant regeneration from seed." *Global Change Biology* 17, 6 (2011): 2145–2161.

Waller, Lauren P., Warwick J. Allen, B. I. P. Barratt, L. M. Condron, Filipe M. França, J. E. Hunt, N. Koele et al. "Biotic interactions drive ecosystem responses to exotic plant invaders." *Science* 368, 6494 (2020): 967–972.

Wang, Hongyan, Long Wang, Mengke Yang, Ning Zhang, Jiazhen Li, Yuqian Wang, Yue Wang, Xuewen Wang, Yanan Ruan, and Sheng Xu. "Growth and DNA methylation alteration in rice (*Oryza sativa* L.) in response to ozone stress." *Genes* 14, 10 (2023): 1888.

Wang, Yongqin, Weihong Dong, Malay C. Saha, Michael K. Udvardi, and Yun Kang. "Improved node culture methods for rapid vegetative propagation of switchgrass (*Panicum virgatum* L.)." *BMC Plant Biology* 21 (2021): 1–12.

Wang, Yuman, Jintong Zhao, Hye Jin Park, and Da Zhou. "Effect of dedifferentiation on noise propagation in cellular hierarchy." *Physical Review E* 105, 5 (2022): 054409.

Wei, Zihan, Quyet Van Le, Wanxi Peng, Yafeng Yang, Han Yang, Haiping Gu, Su Shiung Lam, and Christian Sonne. "A review on phytoremediation of contaminants in air, water and soil." *Journal of Hazardous Materials* 403 (2021): 123658.

Wijerathna-Yapa, Akila, and Jayeni Hiti-Bandaralage. "Tissue culture—a sustainable approach to explore plant stresses." *Life* 13, 3 (2023): 780.

Xu, Fengjiao, Anjali Kariyarath Valappil, Ramya Mathiyalagan, Thi Ngoc Anh Tran, Zelika Mega Ramadhania, Muhammad Awais, and Deok Chun Yang. "In vitro cultivation and ginsenosides accumulation in *Panax ginseng*: a review." *Plants* 12, 17 (2023): 3165.

Xu, Hu, Lixiao Zhang, Kang Zhang, and Yidong Ran. "Progresses, challenges, and prospects of genome editing in soybean (*Glycine max*)." *Frontiers in Plant Science* 11 (2020): 571138.

Yadav, Honey, Kapil Malik, Shashi Kumar, and Pawan K. Jaiwal. "Comparative regeneration in six bread wheat (*Triticum aestivum* L.) varieties from immature and mature scutella for developing efficient and genotype-independent protocol prerequisite for genetic improvement of wheat." *In Vitro Cellular & Developmental Biology-Plant* 56 (2020): 610–617.

Yang, Liwen, Shuangrong Liu, and Rongcheng Lin. "The role of light in regulating seed dormancy and germination." *Journal of Integrative Plant Biology* 62, 9 (2020): 1310–1326.

Yang, Xiyan, and Xianlong Zhang. "Regulation of somatic embryogenesis in higher plants." *Critical Reviews in Plant Science* 29, 1 (2010): 36–57.

Zeng, Yiwen, Sean Maxwell, Rebecca K. Runting, Oscar Venter, James E. M. Watson, and L. Roman Carrasco. "Environmental destruction not avoided with the sustainable development goals." *Nature Sustainability* 3, 10 (2020): 795–798.

Zhen, Nahui, Bojie Fu, Yihe Lu, and Shuai Wang. "Poverty reduction, environmental protection and ecosystem services: a prospective theory for sustainable development." *Chinese Geographical Science* 24 (2014): 83–92.

2 Cellular and Molecular Mechanisms of Plant Regeneration

Akash Deep and Dhananjay Kumar Pandey

2.1 INTRODUCTION

Plants have evolved diverse survival strategies to thrive in harsh environmental conditions, with recovery from damage being a critical aspect of their survival. While animal regeneration primarily focuses on organ or tissue repair post-injury, in plants, regeneration encompasses not only tissue and organ repair but also the formation of entirely new plants (Sugimoto, Gordon, and Meyerowitz 2011; Sena and Birnbaum 2010). For instance, when branches are pruned from a tree trunk, new buds can regenerate. The regenerative capacities of plants have been found to be widespread and have applications in modern agriculture (Sussex 2008). Plant tissue culture, a technique involving the cultivation of plant tissues on a medium supplemented with various plant hormones, harnesses the pluripotency and totipotency of plant cells to facilitate rapid propagation of plant populations. Although plant tissue culture has a rich history (Thorpe 2007; 2012) and is extensively used in agriculture today, there is still much to uncover about the cellular and molecular mechanisms underlying regeneration. Recent years have seen the adoption of molecular genetic approaches to explore plant regeneration, particularly in model plants, resulting in significant advancements in the field. In this chapter, we have focused on the cellular and molecular mechanisms driving plant regeneration, along with exploring several emerging questions pertaining to this process.

2.2 CELLULAR PROCESSES IN PLANT REGENERATION

Plant regeneration can be categorized into various types, each exhibiting a range of forms: (1) restoration of a tissue structure lost due to injury, (2) creation of a new tissue or growth structure not existing before the injury and (3) regeneration of a whole plant from a single somatic cell. In addition to these classical forms of regeneration, plants often regenerate by forming an entire new plant from a cut tissue piece in a controlled environment (Sugimoto, Gordon, and Meyerowitz 2011). In this form of regeneration, entirely new points of plant growth emerge at the site of the injury. These growth points often establish new regions of stem cell activity known as meristems, facilitating ongoing growth and potentially resulting in new individual plants. *De novo* regeneration typically occurs through an intermediate mass of growing cells called calluses. Callus formation can also be induced in cell and tissue cultures by

DOI: 10.1201/9781032702391-2

appropriately adjusting the ratios of the plant hormones auxin and cytokinin. From the growing callus, individual leaves, shoot and root meristems, or newly formed embryos can be cultivated by adjusting the ratios of these plant hormones in the culture medium (Fehér 2019).

2.2.1 Dedifferentiation and Callus Formation in Plants

Historically, dedifferentiation in plant regeneration has been inferred indirectly by observing the renewed capacity of previously inactive cells to undergo division as well as by noting the structural and morphological changes occurring in dividing callus cells. Throughout post-embryonic development, tissues originate from pluripotent stem cells located within specialized structures, known as meristems (Grafi et al. 2011). Cells within meristems are characterized by their small size, dense cytoplasm and small vacuoles. In contrast, quiescent cells within mature tissues, which differentiate from meristems, tend to be larger and contain larger vacuoles, as well as different plastids and storage products, compared to meristematic cells. Callus initiation during regeneration prompts cells to acquire characteristics resembling those of meristematic cells, leading to the loss of differentiated cell morphology. These changes in the cell structure have been interpreted as evidence of dedifferentiation.

However, it is important to note that these changes in cell morphology do not necessarily signify a return to a fundamental embryonic state or to a developmental stage earlier in their usual lineage. The callus, a mass of proliferating cells formed after injury and stimulated in culture by the plant hormones auxin and cytokinin, has long been regarded as dedifferentiated tissue. Nevertheless, recent research has demonstrated that callus is not entirely undifferentiated (Sugimoto, Jiao, and Meyerowitz 2010; Atta et al. 2009). In fact, the callus induced in these studies resembles root tissue in its patterns of gene expression and multicellular organization, even when it is generated from aerial organs that are separated from the root lineage from the earliest divisions of embryogenesis. The transcriptome of calli is more similar to the root meristem tissue than to the shoot meristem or embryonic tissues (Shim et al. 2020).

Therefore, callus formation is not a simple reprogramming process in the dedifferentiated or embryonic state. The regeneration of an entire plant from a single somatic cell could plausibly involve a return to a more embryonic state; therefore, it is easy to accept that dedifferentiation could play a widespread role in various regeneration processes. This assumption is reinforced by the ability of plant cells to undergo somatic embryogenesis in which somatic cells are stimulated to form embryos under *in vitro* culture conditions (Sugimoto, Gordon, and Meyerowitz 2011).

2.2.2 Totipotency and Totipotent Stem Cells in Plants

Plants have demonstrated remarkable regenerative capabilities, as evidenced by their ability to regenerate entire organisms from a wide array of organs. Traditional cuttings and rooting techniques have shown that stems, roots, leaves, hypocotyls, floral axes and even flowers can all be used to regenerate a whole plant. The ease with which plant regeneration can be initiated often leads to the assumption that all plant cells are totipotent. However, this hypothesis remains unclear. The inability to track individual cells from the initiation to the completion of the process in traditional experiments has hindered the validation of this concept (Steward, Ammirato, and Mapes 1970).

Recent evidence contradicts previous findings, indicating that stem cells are responsible for this remarkable regeneration capability. Specifically, the cells from which the callus originates in the roots, hypocotyls, cotyledons and petals are specialized cells located around the vasculature. In roots and hypocotyls, these cells are known as xylem pole pericycle cells (Che, Lall, and Howell 2007). In other organs, the cells that form calluses share the expression of at least one reporter gene with the xylem pole pericycle cells. As these cells have not been previously described in aerial organs, there is presently no standardized nomenclature for them. Similar to mammalian stem cell niches often found around the vasculature, these plant pericycle-like cells seem to surround the vasculature of various organ types and serve as the source of regenerating tissues (Müller-Xing and Xing 2022).

Although the role of pericycle cells in regeneration has only recently been established, these cells have long been recognized as the stem cells responsible for lateral root formation (Y. Zhang, Umeda, and Kakimoto 2022). During lateral root formation, a restricted number of pericycle cells positioned adjacent to the two xylem poles undergo organized cell division and differentiation. These cells generate all the cell layers found in the root, as they give rise to a lateral root primordium (LRP) (Kurup et al. 2005). Similar to animals, the regulation of this stem cell niche in plants is influenced by diffusible substances originating from other cells. Hormones have been demonstrated to play a role in the formation of the LRP from the pericycle cells. Auxin positively regulates the initiation and development of each LRP. Local accumulation of auxin in pericycle cells serves as the initial trigger for the LRP formation (Dubrovsky et al. 2008). However cytokinin negatively regulates LRP formation by modifying the expression of auxin transporter PIN proteins. This alteration disrupts the establishment of an auxin gradient, thereby permitting root reorganization (Laplaze et al. 2007). Likewise, the induction of callus formation in culture necessitates supplementation with exogenous hormones in the medium.

Interestingly, callus formation mirrors the developmental program of lateral root initiation at its onset, even when it is derived from aerial organs. This correlation is evidenced by the observation that a mutant with impaired lateral root formation also displays deficiencies in callus formation (Atta et al. 2009). Hence, in this system, the differentiation of pericycle-like cells into tissues resembling the root meristem is the shared mechanism underlying callus formation across different organs.

Additionally, recent findings indicate that root pericycle cells possess the ability to directly generate shoots in locations where lateral root primordia would typically develop when exposed to cytokinin-rich media (Atta et al. 2009). Thus, in plants, the regeneration of at least some tissues involves distinct populations of cells that can be considered adult stem cells. These cells contribute to regenerated material by differentiating from stem cells of one tissue type or potentially transdifferentiating to a completely different stem cell (Shen, Burke, and Tosh 2004). Unlike in vertebrate animals, in plants, one common type of adult stem cell is distributed throughout the entire body along the vasculature. This accessibility of the adult stem cells might be one of the reasons for the high plasticity of plants, relative to vertebrates (Sugimoto, Gordon, and Meyerowitz 2011).

2.3 MOLECULAR MECHANISMS OF PLANT REGENERATION

2.3.1 Role of Hormonal Regulation in Plant Regeneration

Auxins and cytokinins are commonly used to induce callus formation. However, the precise molecular mechanisms underlying this induction remain relatively understudied. Recent research has shown that several regulators involved in lateral root development play a role in callus formation in callus-inducing medium (CIM) (Ikeuchi, Sugimoto, and Iwase 2013). Auxin is one of the most common plant hormones that can accumulate and trigger responses within cells and tissues. Localized accumulation of auxin and its response dynamics are frequently observed through reporters such as green fluorescent protein (GFP) and yellow fluorescent protein (YFP), which are controlled by the auxin-responsive DR5 element (Liao et al. 2015). Cytokinin, another crucial hormone governing plant growth and development, plays a significant role.

Plasmodesmata (PD) serve as gateways within plant cell walls, which undergo precise regulation of permeability throughout the stages of plant growth and development. The involvement of the actin cytoskeleton has been identified in the regulation of PD permeability. A variety of actin-binding proteins (ABPs), including members of the gelsolin/villin/fragmin superfamily and nucleating proteins such as formin, profilin and cofilin, play crucial roles in orchestrating cellular and cytoskeletal dynamics (Diao and Huang 2021; Deep and Pandey 2024; Pandey and Chaudhary 2015; 2024; 2021). These structures are pivotal in various aspects of plant development and are integral to coordinating cellular and systemic signaling responses when plants interact with both biotic and abiotic factors in their environment.

Researchers have assessed the involvement of cytokinin responses in plant organ development by using reporters driven by TCS (two-component signaling sensor), a synthetic promoter responsible for cytokinin activity (Kirschner et al. 2018). Asymmetric auxin response signals play vital roles in various aspects of plant growth and development, including embryo axis formation, flower primordium initiation and patterning, vascular tissue differentiation, root meristem maintenance and tropical growth (Schepetilnikov and Ryabova 2017). Previous studies have explored the involvement of cytokinins in the regeneration of cultured plant tissue. During callus formation in root explants, cytokinin-responsive signals visualized by *ARABIDOPSIS RESPONSE REGULATORS (ARR5)* are primarily observed in the vasculature of root explants. These signals were notably concentrated in proliferating callus cells cultivated on CIM.

Although the patterning of auxin response in plant development is well established, its involvement in the regeneration of plant tissues in culture remains poorly understood. The distribution of auxin-responsive signals was examined during callus formation in *Arabidopsis* root explants cultured on CIM, which are rich in auxins (Gordon et al. 2007). Notably, areas where shoot apical meristem (SAM) initiation occurs exhibit low or undetectable levels of auxin-responsive signals, in contrast to the strong presence of these signals in surrounding regions. The spatial patterns of auxin responses are also distinctly evident during shoot regeneration induced by pistils, as shown in Figure 2.1 (Cheng et al. 2012; Bhojwani and Razdan 1996).

Uniform detection of auxin-responsive signals occurs at the periphery of mature calli in CIM. Following shoot induction on SIM, these signals migrate to

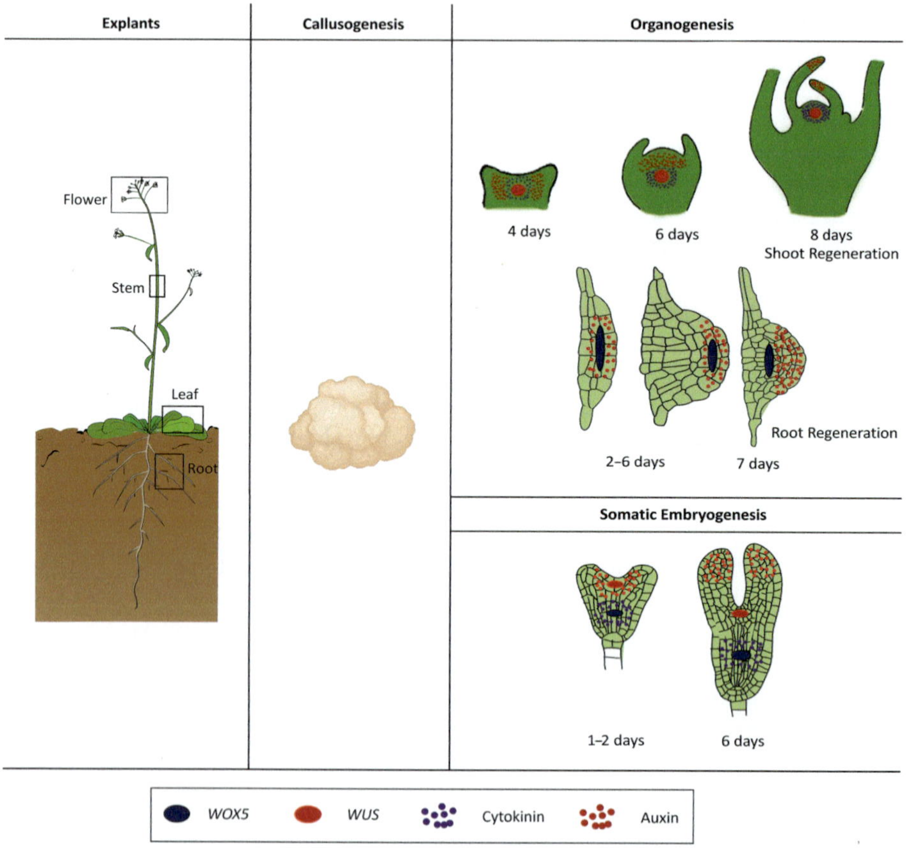

FIGURE 2.1 Patterns of hormone responsiveness in shoot, root and somatic embryogenesis (SE) regeneration. Explants are excised and transferred to CIM to stimulate callus formation (Cheng et al. 2012). Calli cultured on CIM for 20 days are then moved to SIM for shoot induction. *WUS* expression initiates at four days in the organizing center (OC) of the developing shoot apical meristem (SAM). At this stage, auxin response is detected in the regions surrounding *WUS* expression, while cytokinin-responsive signals are concentrated in the central region of the SAM, coinciding with *WUS* expression. By six days, when the shoot primordium emerges, both auxin and cytokinin responses are observed at the apex of the primordium. During the initiation of root tissue, auxin response displays a spatial distribution pattern for root induction on RIM around two days, correlating with the expression domain of *WOX5*. Subsequently, *WOX5* expression becomes localized in the QC of the regenerated roots, with auxin-responsive signals present in the root apex. Embryonic callus is cultured on auxin-rich ECIM for 14 days. Following SE induction on SEIM, auxin-responsive gradients are established in the peripheral regions surrounding areas of *WUS* expression, marking the OC of promeristems (Su et al. 2009; 2010). Later, auxin-responsive signals are reorganized at the apex of the promeristems for the initiation of somatic proembryos. Cytokinin-response signals are distributed asymmetrically across the regions of *WOX5* expression.

the outermost cell layer regions, as depicted in Figure 2.1, surrounding the domain expressing *WUS* (Su and Zhang 2014). Endogenous auxin-responsive signals were absent from the edges of the embryonic calli when cultured on auxin-rich embryonic callus-inducing medium (ECIM). However, upon removal of exogenous auxin from the ECIM to induce somatic embryogenesis, endogenous auxin-responsive gradients form in the edge regions surrounding areas expressing *WUS*. This expression marks the organizing centers (OCs) of promeristems that eventually develop into somatic proembryos in somatic embryo-inducing medium (SEIM), as shown in Figure 2.1 (Su et al. 2009; 2010). After root induction on RIM for two to four days, a limited distribution of auxin-responsive signals was observed, as depicted in Figure 2.1, consistent with the expression patterns of the root meristem-specific gene *WUSCHEL-RELATED-HOMEOBOX 5 (WOX5)* (Savina et al. 2020).

In mature calli derived from pistils cultured on CIM without shoot induction, the cytokinin response was observed mainly in the edge regions, mirroring the pattern of the auxin response. Upon *de novo* shoot initiation in SIM, there is a regional shift in cytokinin-responsive signals toward areas where SAM initiation and *WUS* expression occur. Additionally, the spatial expression pattern of the cytokinin-responsive gene *ARR5* in *in vitro*-induced root organs suggests that root regeneration involves a localized cytokinin response within the callus (Pernisová et al. 2009). Notably, unlike the distribution of auxin-responsive signals in somatic embryogenesis promeristems, cytokinin-response signals are unevenly distributed in areas expressing *WOX5*, associated with the embryonic root meristem (Lee et al. 2022). These findings suggest that cytokinins play a significant role in the establishment of the root apical meristem during somatic embryogenesis.

2.3.2 Molecular Pathways, Signaling, Genes and Transcription Factors in Plant Regeneration

Plants have developed diverse survival strategies to thrive under harsh environmental conditions, and recovery from damage is a crucial aspect of survival. While animals primarily require regeneration to repair organs or tissues following injury, plant regeneration encompasses not only tissue and organ repair but also the generation of new plants. The regenerative capacity of plants has been exploited extensively in modern agriculture. (Sussex 2008; Sugimoto, Gordon, and Meyerowitz 2011).

Regeneration of higher plants can be divided into three primary classifications based on the genes and transcription factors involved in tissue regeneration, *de novo* organogenesis and somatic embryogenesis.

2.3.2.1 Tissue Regeneration

At the tip of the root apical meristem (RAM), the root stem cell niche consists of a quiescent center (QC) surrounded by the stem cells. Typically, QC comprises two to four cells that divide at a very slow rate and are characterized and regulated by the transcription factor gene *WUSCHEL-RELATED HOMEOBOX 5 (WOX5)* (Strotmann and Stahl 2021). Recent investigations of *Arabidopsis* root tips have revealed the molecular mechanisms underlying root tip regeneration (Savina et al. 2020). Following laser-induced ablation of QC cells, notable alterations were observed as shown in in

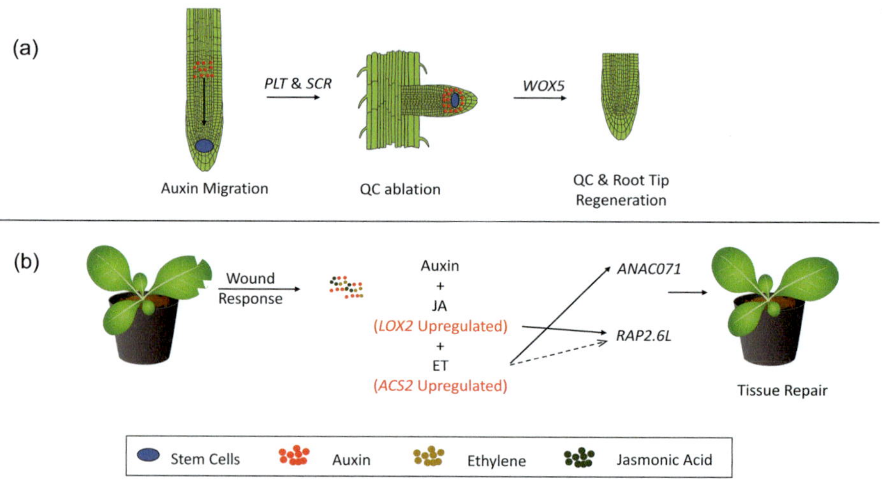

FIGURE 2.2 In *Arabidopsis*, tissue regeneration processes are observed in various contexts. (a) Regeneration of root tips occurs following the ablation of quiescent center (QC) cells. (b) Tissue repair mechanisms injury disrupts the basipetal flux of auxin, resulting in differential auxin concentrations between the upper and lower regions of the wound site. This disparity in auxin levels initiates the regeneration process to facilitate wound repair.

auxin polar transport and expression levels of crucial transcription factors during the regeneration process of the new QC, as shown in Figure 2.2a. Elevated auxin levels result in the upregulation of *PLETHORA (PLT)* genes, which facilitates the nuclear localization of SHORTROOT (SHR) proteins, thereby activating *SCARECROW (SCR)* gene. As depicted in Figure 2.2a, collaboration between *PLTs* and *SCR* subsequently prompts the transformation of cell fates in QC-adjacent cells into new QC cells by reorganizing PIN proteins and reactivating *WOX5* expression in the newly formed QC (Matosevich and Efroni 2021). Hence, it seems that the entire process of QC formation is primarily governed by the level of auxin, with a specifically high concentration of auxin being a pivotal factor in determining the fate of QC cells.

Sena et al. employed a RAM excision method to investigate the process of root regeneration (Sena and Birnbaum 2010). The distinction between QC ablation and RAM excision lies in the fact that ablation specifically targets QC for destruction, whereas RAM excision involves removal of the entire root apical meristem (Xu et al. 2006), whereas the entire stem cell niche is removed by excision. Hence, the excision experiment aimed to ascertain whether a stem cell niche was necessary for the regeneration of a new QC. The roots from which the RAM was removed were capable of generating a new QC. Compared with the QC regeneration observed in the ablation experiment, the newly regenerated QC was governed by the accumulation of elevated levels of auxin, as shown in Figure 2.2a (Sena and Birnbaum 2010).

Recent studies utilizing incision experiments have elucidated the mechanisms by which plant tissues undergo self-repair following injury. Upon stem injury, the basipetal auxin stream is obstructed at the wound site, leading to the accumulation of high auxin levels above the wound site and relatively low auxin levels below it. Elevated auxin concentrations trigger the expression of NAC (NAM, ATAF1,2 and

CUC2) family transcription factor gene *ANAC071*, whereas diminished auxin levels promote the expression of APETALA2 (AP2)/ERF family transcription factor gene *RAP2.6L*. As discussed in Figure 2.2b, both genes play critical roles in wound healing; however, the precise molecular functions of these transcription factors remain unclear. The lipoxygenase gene *LOX2* is upregulated below the incision in a pattern similar to that of RAP2.6L. Additionally, the application of methyl jasmonate increases *RAP2.6L* expression. Ethylene levels are inferred from the expression of *ACS2* gene, which belongs to a family of aminocyclopropane carboxylic acid synthase genes that regulate the rate-limiting step in ethylene (ET) biosynthesis, rather than being directly measured (Reid and Ross 2011). In addition to the pivotal role of auxins, wound-induced jasmonic acid (JA) and ethylene production are necessary to regulate *RAP2.6L* (Asahina et al. 2011).

2.3.2.2 *De novo* Organogenesis

In 1950s, Skoog and Miller formulated a medium that was suitable for both callus induction and *de novo* organogenesis and proposed that shoot or root initiation in tissue culture mainly relies on different ratios of auxin and cytokinin in the medium (Skoog and Miller 1957). Callus induction is a widely used method for plant regeneration. Balancing hormone levels and coordinating interactions between auxins and cytokinins are essential for determining cell fate and organizing cell patterns during *de novo* organogenesis (Cheng et al. 2012).

Elevated levels of cytokinin promote the regeneration of shoots from callus cells, as evidenced by the role of cytokinins in disrupting early root identity, causing disarray in root primordia and triggering the onset of shoot development (Pernisová et al. 2009). Several components of cytokinin biosynthesis and signaling are pivotal for *de novo* shoot organogenesis. For instance, activation of the *LONELY GUY (LOG)* gene, which is responsible for cytokinin biosynthesis, during plant regeneration leads to cell cycle re-entry at the G_1-S cell cycle phase by inducing *CYCD3* expression (Ikeuchi et al. 2017; Riou-Khamlichi et al. 1999).

Adenosine phosphate-isopentenyltransferases (IPTs), which catalyze a key step in cytokinin biosynthesis, also promote spontaneous shoot formation from calli, even in the absence of cytokinins (Kakimoto 2001). Additionally, histidine kinases such as AHK2, AHK3 and AHK4/WOODENLEG (WOL)/CYTOKININ RESPONSE 1 (CRE1) in *Arabidopsis* play a positive role in *de novo* shoot organogenesis (Lardon et al. 2022). These kinases are likely to enhance *WUS* expression while suppressing *WOX5* expression to facilitate shoot formation (Pernisova et al. 2018). Mutations in *AHK2*, *AHK3* and *AHK4* diminish the ability of calli to regenerate shoots. Consequently, modulation of cytokinin signaling downstream of *Arabidopsis* histidine phosphotransfer proteins (AHPs) significantly influences shoot regeneration (Lardon et al. 2020).

Auxin is a crucial hormone that influences *de novo* shoot organogenesis. *AUXIN RESPONSE FACTORs* (*ARFs*) serve as pivotal mediators of auxin signaling, predominantly contributing to plant regeneration (Pandey and Chaudhary 2015). For instance, *MONOPTEROS* (*MP*)/*ARF5* plays a role in shoot formation from calli, potentially by activating the transcription of *STM* and *CYTOKININ RESPONSE FACTOR 2 (CRF2)* (Ckurshumova et al. 2014). Similarly, calli derived from pistils of *arf3* mutant plants exhibited a notable decrease in shoot regeneration in SIM (Cheng et al. 2012). Furthermore, the miRNA-mediated control of ARFs adds another regulatory dimension to shoot regeneration. miR160 targets *ARF10* transcripts post-transcriptionally and has a positive effect on *de novo* shoot organogenesis. Elevated miR160 levels impede shoot regeneration. Conversely, explants expressing a miR160-resistant variant of *ARF10* demonstrated enhanced shoot regeneration compared with wild-type explants, along with increased expression of *CLAVATA 3 (CLV3)*, *CUC* genes and *WUS* (Qiao et al. 2012). Since miR160 not only targets *ARF10* but also activates the cytokinin signaling gene *ARR15*, it is conceivable that miR160 orchestrates the crosstalk between auxin and cytokinin signaling (Liu et al. 2016).

Ethylene also plays a role in regulating *de novo* shoot organogenesis, likely through its interaction with the auxin and cytokinin signaling pathways. Ethylene can have both positive and negative effects on the shoot regeneration depending on the sensitivity of explants to ethylene signaling. In *Arabidopsis*, mutants with ethylene overproduction (*eto1-1*) and constitutive ethylene responses (*ctr1-1* and *ctr1–12*) typically display enhanced shoot regeneration. Conversely, mutants insensitive to ethylene (*etr1-1*, *ein2-1*, *ein4* and *ein7*) exhibit reduced rates of *de novo* organogenesis. These effects may stem from the interplay between ethylene, auxin and cytokinin signaling pathways. Supporting this notion, *HOOKLESS 1 (HLS1)*, an ethylene-inducible gene, likely participates in the crosstalk between ethylene and auxin during *de novo* organogenesis, as evidenced by the enhanced shoot formation observed in the *hls1-1* mutant, which displays altered responses to auxin (Chatfield and Raizada 2008).

2.3.2.3 Somatic Embryogenesis

LEAFY COTYLEDON (LEC) and *AGAMOUS-LIKE15 (AGL15)* transcription factors are crucial for the initiation of somatic embryo development in *Arabidopsis,* as shown in Figure 2.3. The *LEC* genes consist of *LEC1*, which encodes the HAP3 subunit of the CCAAT-binding transcription factor, as well as *LEC2* and *FUSCA3 (FUS3)*, which encode B3-domain proteins. *AGL15* encodes an MADS-box transcription factor that is expressed exclusively in embryos (Liu et al. 2021). *LEC* and *AGL15* can enhance each other's expression and regulate embryo cell fate by interacting with hormone pathways (Joshi et al. 2022). Molecular analyses have indicated that *AGL15* suppresses the gibberellic acid (GA) pathway by enhancing the expression of *GA2ox6* and inhibiting *GA3ox2* expression, while also elevating the expression of the auxin signaling gene *IAA30* (Zheng et al. 2009). *FUS3* suppresses the GA pathway by inhibiting the expression of GA biosynthesis genes *GA3ox1* and *GA3ox2*. Furthermore, the *FUS3* and abscisic acid (ABA) action pathways mutually enhance each other, although the exact molecular mechanism remains unclear (Sun 2008). *LEC2* enhances the auxin pathway by increasing the expression of the auxin biosynthesis genes *YUC2* and *YUC4*, as well as the *IAA30* gene.

Together, these transcription factors constitute a genetic network that promotes the ABA and auxin pathways, while suppressing the GA pathway (Wójcikowska et al. 2013). Conversely, *PGA37/MYB118* and *MYB115* have been shown to enhance the expression of *LEC1* (Wang et al. 2009). Increased expression of *BABY BOOM (BBM)* and *EMBRYOMAKER (EMK)*, both of which encode AP2-domain transcription factors, lead to the induction of somatic embryogenesis. *ABI3*, which

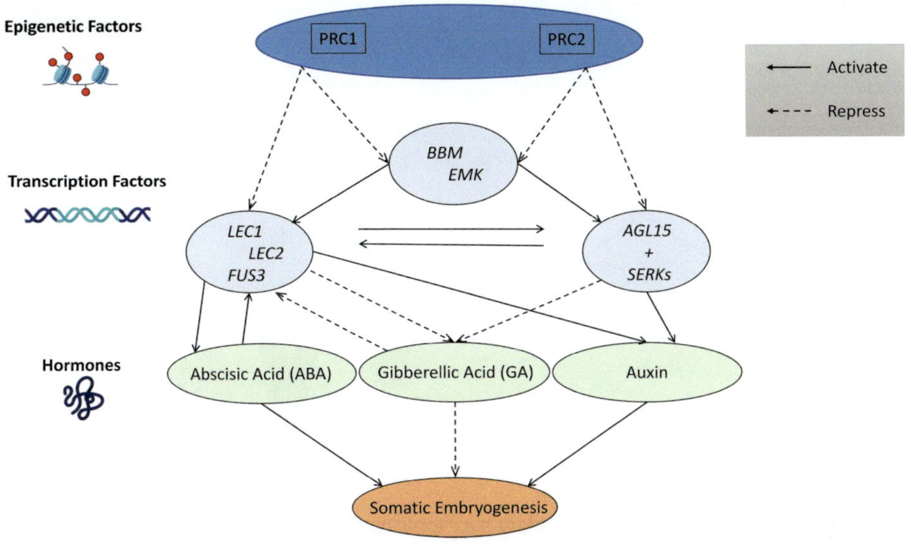

FIGURE 2.3 Somatic embryogenesis, the process of generating embryos from somatic cells, is regulated by a complex interplay of epigenetic mechanisms, transcription factors and plant hormones. A network of plant hormones and transcription factors engages in communication, while epigenetic pathways, such as PRC1 and PRC2, actively suppress the expression of embryo-specific genes (*LEC1*, *LEC2* and *FUS3*) to maintain the fidelity of somatic cells and prevent them from acquiring embryo-like traits.

encodes a B3 domain transcription factor, is positively regulated by *LECs* and *AGL15* (Horstman et al. 2017; Jha and Kumar 2018). A dual mutation affecting the B3 domain genes *VP1/ABI3-LIKE1 (VAL1)* and *VAL2* led to somatic embryogenesis accompanied by abnormal expressions of *LEC1*, *ABI3* and *FUS3*, indicating that *VAL1* and *VAL2* act as inhibitors of somatic embryogenesis (Suzuki, Wang, and McCarty 2007). *WUS* and other genes related to the SAM have been demonstrated to contribute to the promotion of somatic embryogenesis (Su et al. 2009). Moreover, *SOMATIC EMBRYOGENESIS RECEPTOR-LIKE KINASEs (SERKs)* can participate in signaling pathways during somatic embryogenesis by forming a protein complex with *AGL15* (Karlova et al. 2006). The regulatory sequences of *SERK* genes have evolved with distinctive *AINTEGUMENTA* binding sites, resembling *AP2*-like ethylene-responsive transcription factors, to determine their spatial expression in somatic embryogenesis (Pandey and Chaudhary 2014). *SERK1*, implicated in potential signaling pathways during somatic embryogenesis (SE), exhibited notable upregulation in the embryogenic calli of the fruit line (Pandey and Chaudhary 2014).

2.3.3 Role of Micro-RNA in Plant Regeneration

Micro-RNAs, which are 21-nt non-coding small RNAs, play a role in decreasing gene transcription by interacting with mature mRNAs (Axtell 2013). Among them, miR156, miR160, miR167, miR319 and miR393 have been implicated in plant regeneration by modulating auxin and CK signaling genes, either directly or indirectly.

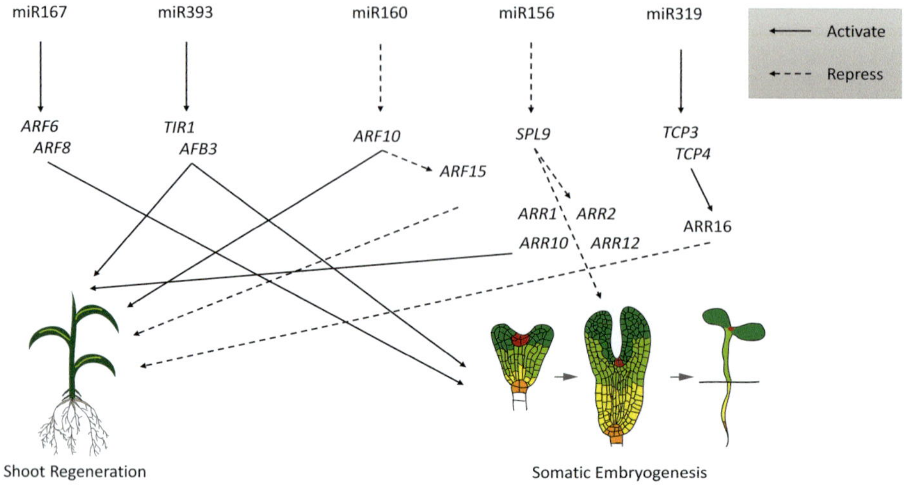

FIGURE 2.4 miRNA regulation of plant regeneration.

miR156 participates in various developmental processes that are associated with age (Guo et al. 2017; Xu et al. 2016). The ability of *Arabidopsis* and tobacco to regenerate shoots decreases as plants age, a decline that can be offset by overexpressing *miR156*. As shown in Figure 2.4, SQUAMOSA PROMOTER BINDING PROTEIN-LIKE 9 (SPL9), a protein encoded by a gene targeted by *miR156*, directly interacts with type-B ARR genes, such as *ARR1*, *ARR2*, *ARR10* and *ARR12*, thereby hindering cytokinin responses. Elevated levels of miR156 suppressed *SPL9* expression during the juvenile phase of *Arabidopsis* seedlings. However, as plants transition from juveniles to adults, reduced miR156 levels result in increased *SPL9* expression, leading to the suppression of cytokinin responses and consequently diminishing shoot regeneration capacity (Zhang et al. 2015).

The miR156-SPL regulatory pathway plays a similar role in the SE of citrus plants. In many citrus cultivars, the ability of calluses to undergo embryogenesis gradually diminishes over prolonged culture periods, resulting in the failure of shoot differentiation. Non-embryonic calluses exhibited significantly lower levels of miR156 than embryonic calli, while the target genes *CsSPL3* and *CsSPL14* showed the opposite trend. The expression levels of *CsSPL3* and *CsSPL14* were strongly inversely correlated with SE capacity across various citrus varieties. In orange varieties, such as 'Anliu,' 'Newhall,' 'Valencia' and' American' sour orange, which demonstrate SE capability, the expression levels of *CsSPL3* and *CsSPL14* are relatively low. Conversely, in varieties with limited competence in SE, the expression levels of *CsSPL3* and *CsSPL14* were high (Long et al. 2018). These findings indicated that the miR156-SPL regulatory mechanism participates in the regulation of age-related and cultivar-specific SE in citrus plants.

Similar to miR156, miR319 facilitates shoot regeneration by modulating cytokinin response. MiR319 targets *the TCP3* and *TCP4* genes, which encode proteins that directly activate ARR16, a negative regulator of shoot regeneration, as shown in Figure 2.4. Reduced function of HUA ENHANCER 1 (HEN1), a small RNA methyltransferase responsible for stabilizing miR319, results in decreased miR319 expression. This decrease leads to the upregulation of *TCP3* and *TCP4*, subsequently activating *ARR16* and suppressing shoot regeneration (Yang et al. 2020).

miRNAs also play a role in modulating auxin signaling during plant regeneration. For instance, miR160 targets *ARF10*, leading to the inhibition of auxin signaling, which subsequently hampers callus and shoot regeneration. Additionally, the transcriptional repressor *ARF10* directly binds to AuxRE in the promoter region of *ARR15*, which encodes a negative regulator of callus formation, as shown in Figure 2.4. Thus, miR160 hinders regeneration by influencing both the auxin and cytokinin signaling pathways. In cotton, miR167 negatively regulated somatic embryogenesis by targeting *ARF6* and *ARF8*. Overexpression of the miR167 target mimic *(35S::MIM167)* in plants results in a significant upregulation of *ARF6* and *ARF8*, the auxin-responsive gene *GRETCHEN HAGEN 3 (GH3)* and the auxin transporter genes *AUXIN RESISTANT 1 (AUX1), LIKE AUX1 3 (LAX3), PIN1* and *PIN2*. This suggests that miR167 promotes somatic embryogenesis by enhancing auxin signaling (Arora, Singh, and Chaudhary 2020). In addition to influencing *ARF* expression, miRNAs also affect the expression of genes encoding auxin receptors, such as *TRANSPORT INHIBITOR RESPONSE 1 (TIR1)* and *AUXIN SIGNALING F-BOX 3 (AFB3)*. In the miR393 mutant, *TIR1* and *AFB3* were observed to be upregulated, resulting in an increased capacity for shoot regeneration and somatic embryogenesis compared to the wild type, as shown in Figure 2.4 (Wang et al. 2018).

2.4 FACTORS INFLUENCING PLANT REGENERATION

2.4.1 Epigenetic Control of Regeneration

Plant cell regeneration is usually triggered by damage, but recent studies suggest that multiple epigenetic mechanisms actively suppress this regenerative ability during normal development (Ikeuchi, Iwase, and Sugimoto 2015). One such mechanism involves POLYCOMB REPRESSIVE COMPLEXES 1 and 2 (PRC1 and PRC2) are chromatin modifiers that sustain transcriptional repression by catalyzing the trimethylation of histone H3 at lysine 27 (H3K27me3). (Holec and Berger 2012). A study demonstrated that plants with mutations in *PRC2* initially develop roots similar to wild-type plants, with fully formed root hair cells, but later undergo reprogramming to form calli and embryo-like structures (Ikeuchi et al. 2015). In this context, as depicted in Figure 2.3, PRC2 suppresses key regulators of reprogramming, such as *WIND3*, and embryonic regulators, such as *LEC2*, leading to elevated expression of these genes and subsequent cellular reprogramming. Additionally, histone deacetylation, another mechanism implicated in transcriptional repression, may act as a safeguard against premature somatic embryogenesis. When wild-type *Arabidopsis* plants are treated with trichostatin A (TSA), a histone deacetylase inhibitor, embryo-like structures form from the true leaves (Tanaka, Kikuchi, and Kamada 2008).

Similarly, mutations that cause loss of function in two histone deacetylases, HDA19 and HDA6, lead to the formation of embryo-like structures in shoots (Tanaka, Kikuchi, and Kamada 2008). This phenomenon is linked to the abnormal expression of key embryonic regulators like *LEC1* and *LEC2* and can be counteracted by introducing the *lec1* mutation (Tanaka, Kikuchi, and Kamada 2008). Interestingly, combining TSA treatment with heat stress significantly boosted the efficiency of somatic embryogenesis in *Brassica napus* microspores (Li et al. 2014). This implies that heat stress and histone deacetylation might converge to promote the upregulation of embryonic regulators, initiating the embryonic program. Additionally, genetic studies in *Arabidopsis* suggest that other epigenetic mechanisms also play a role in regulating organ regeneration. Mutations in *DNA METHYLTRANSFERASE1 (MET1)* boost shoot regeneration in SIM, along with elevated expression of several *MET1* targets, such as *WUS* (Li et al. 2011). Like many other regeneration regulators, the *WUS* locus exhibits various epigenetic marks that change when *WUS* expression rises during shoot regeneration (Li et al. 2011). Clarifying the causal relationships between these epigenetic modifications and changes in gene expression is a crucial objective for future research.

2.4.2 Environmental Constraints that Impact Plant Regeneration

The regenerative ability of plant explants is influenced by various environmental factors, including nutrient composition, gelling agents, pH, light and temperature (George, Hall, and De Klerk 2007). Light exposure is a well-known environmental factor that affects plant regeneration; however, its effects seem to vary depending on the context. In some plant species, light is necessary for shoot regeneration (Reuveni and Evenor 2007) and can stimulate organ regeneration (Gutierrez et al. 2009). However, in certain situations, exposure to light can inhibit root or shoot regeneration (Bellini, Pacurar, and Perrone 2014).

Tissue culture experiments with *Arabidopsis* cotyledons have revealed that initial exposure to light shortly after excision is generally detrimental to shoot regeneration, and even brief periods of darkness, ranging from 2 to 6 hours, can enhance regeneration (Nameth et al. 2013). Light exposure triggers multiple signaling pathways, some of which lead to oxidative damage through ROS generation of reactive oxygen species. Two photoreceptors are involved in the light response during shoot regeneration: CRYPTOCHROME1 (CRY1), which is a blue/UV-A light receptor that significantly inhibits shoot regeneration, and PHYTOCHROME A (PHYA), which is a far-red light receptor that shields explants from initial light inhibition (Nameth et al. 2013). A key downstream regulator in light signaling is the transcription factor ELONGATED HYPOCOTYL5 (HY5), which seems to protect explants from light exposure by promoting anthocyanin accumulation. *Arabidopsis* accessions show diverse responses to light during shoot regeneration, indicating that genetic variations in this process present an interesting avenue for future research.

2.5 CONCLUSION AND FUTURE PERSPECTIVES

During regeneration, certain intrinsic developmental processes are activated in response to external stimuli, integrating developmental and environmental signals, which leads to varying regeneration strategies and efficiencies. Since regeneration originates from a small cell population within somatic tissues, identifying these cells and understanding how external stresses influence cell fate is essential. In normal development, many key regulators of regeneration are epigenetically silenced to prevent inappropriate cellular reprogramming, and overcoming these repressions through external stimuli remains a major challenge.

Research in *Arabidopsis* has shed light on how plants regenerate from relatively undifferentiated cells, but different mechanisms may be at play in plants regenerating from differentiated cells. With the rapid advancements in next-generation sequencing and genome editing technologies, we can now explore the molecular mechanisms behind these lesser-known forms of regeneration and conduct functional studies in non-model plants. QTL analysis of accessions with varying regeneration efficiencies has proven to be a valuable complementary approach, and identifying new QTLs should help uncover novel regeneration mechanisms.

A key goal in plant regeneration research is to use our understanding of basic biology to develop new molecular tools that enhance regeneration efficiency in crops. For example, the ectopic expression of key regeneration regulators like *WUS* and *WIND1* has been shown to promote organ regeneration and somatic embryogenesis in various crops. Additionally, the expression profiles of these regulators have been used to identify cultivars with high regeneration capacities. Further insights into plant regeneration mechanisms should advance tissue culture techniques with significant implications for both fundamental and applied biology.

REFERENCES

Arora, Sakshi, Amarjeet Kumar Singh, and Bhupendra Chaudhary. 2020. 'Target-Mimicry Based miRNA167-Diminution Ameliorates Cotton Somatic Embryogenesis via Transcriptional Biases of Auxin Signaling Associated miRNAs and Genes'. *Plant Cell, Tissue and Organ Culture (PCTOC)* 141 (3): 511–31. https://doi.org/10.1007/s11240-020-01810-9.

Asahina, Masashi, Katsuya Azuma, Weerasak Pitaksaringkarn, Takashi Yamazaki, Nobutaka Mitsuda, Masaru Ohme-Takagi, Shinjiro Yamaguchi, et al. 2011. 'Spatially Selective Hormonal Control of RAP2.6L and ANAC071 Transcription Factors Involved in Tissue Reunion in *Arabidopsis*'. *Proceedings of the National Academy of Sciences of the United States of America* 108 (38): 16128–32. https://doi.org/10.1073/pnas.1110443108.

Atta, Ramzy, Lieve Laurens, Elodie Boucheron-Dubuisson, Anne Guivarc'h, Eugénie Carnero, Véronique Giraudat-Pautot, Philippe Rech, and Dominique Chriqui. 2009. 'Pluripotency of Arabidopsis Xylem Pericycle Underlies Shoot Regeneration from Root and Hypocotyl Explants Grown *In Vitro*'. *The Plant Journal: For Cell and Molecular Biology* 57 (4): 626–44. https://doi.org/10.1111/j.1365-313X.2008.03715.x.

Axtell, Michael J. 2013. 'Classification and Comparison of Small RNAs from Plants'. *Annual Review of Plant Biology* 64 (1): 137–59. https://doi.org/10.1146/annurev-arplant-050312-120043.

Bellini, Catherine, Daniel I. Pacurar, and Irene Perrone. 2014. 'Adventitious Roots and Lateral Roots: Similarities and Differences'. *Annual Review of Plant Biology* 65 (1): 639–66. https://doi.org/10.1146/annurev-arplant-050213-035645.

Bhojwani, S. S., and M. K. Razdan, eds. 1996. 'Production of Pathogen-Free Plants'. In *Studies in Plant Science*, 5: 451–82. Plant Tissue Culture. Elsevier. https://doi.org/10.1016/S0928-3420(96)80017-6.

Chatfield, Steven P., and Manish N. Raizada. 2008. 'Ethylene and Shoot Regeneration: *Hookless1* Modulates de Novo Shoot Organogenesis in *Arabidopsis Thaliana*'. *Plant Cell Reports* 27 (4): 655–66. https://doi.org/10.1007/s00299-007-0496-3.

Che, Ping, Sonia Lall, and Stephen H. Howell. 2007. 'Developmental Steps in Acquiring Competence for Shoot Development in *Arabidopsis* Tissue Culture'. *Planta* 226 (5): 1183–94. https://doi.org/10.1007/s00425-007-0565-4.

Cheng, Zhi Juan, Liang Wang, Wei Sun, Yan Zhang, Chao Zhou, Ying Hua Su, Wei Li, et al. 2012. 'Pattern of Auxin and Cytokinin Responses for Shoot Meristem Induction Results from the Regulation of Cytokinin Biosynthesis by AUXIN RESPONSE FACTOR3'. *Plant Physiology* 161 (1): 240–51. https://doi.org/10.1104/pp.112.203166.

Ckurshumova, Wenzislava, Tatiana Smirnova, Danielle Marcos, Yara Zayed, and Thomas Berleth. 2014. 'Irrepressible *MONOPTEROS/ARF 5* Promotes *de Novo* Shoot Formation'. *New Phytologist* 204 (3): 556–66. https://doi.org/10.1111/nph.13014.

Deep, Akash, and Dhananjay K. Pandey. 2024. 'Genome-Wide Analysis of *VILLIN* Gene Family Associated with Stress Responses in Cotton (*Gossypium* Spp.)'. *Current Issues in Molecular Biology* 46 (3): 2278–300. https://doi.org/10.3390/cimb46030146.

Diao, Min, and Shanjin Huang. 2021. 'An Update on the Role of the Actin Cytoskeleton in Plasmodesmata: A Focus on Formins'. *Frontiers in Plant Science* 12 (February). https://doi.org/10.3389/fpls.2021.647123.

Dubrovsky, Joseph G., Michael Sauer, Selene Napsucialy-Mendivil, Maria G. Ivanchenko, Jiří Friml, Svetlana Shishkova, John Celenza, and Eva Benková. 2008. 'Auxin Acts as a Local Morphogenetic Trigger to Specify Lateral Root Founder Cells'. *Proceedings of the National Academy of Sciences of the United States of America* 105 (25): 8790–94. https://doi.org/10.1073/pnas.0712307105.

Fehér, Attila. 2019. 'Callus, Dedifferentiation, Totipotency, Somatic Embryogenesis: What These Terms Mean in the Era of Molecular Plant Biology?' *Frontiers in Plant Science* 10 (April). https://doi.org/10.3389/fpls.2019.00536.

George, Edwin F., Michael A. Hall, and Geert-Jan De Klerk. 2007. *Plant Propagation by Tissue Culture: Volume* 1. *The Background*. Vol. 1. Dordrecht: Springer Science & Business Media. eBook ISBN: 978-1-4020-5005-3

Gordon, Sean P., Marcus G. Heisler, G. Venugopala Reddy, Carolyn Ohno, Pradeep Das, and Elliot M. Meyerowitz. 2007. 'Pattern Formation during de Novo Assembly of the *Arabidopsis*shoot Meristem'. *Development* 134 (19): 3539–48. https://doi.org/10.1242/dev.010298.

Grafi, Gideon, Vered Chalifa-Caspi, Tal Nagar, Inbar Plaschkes, Simon Barak, and Vanessa Ransbotyn. 2011. 'Plant Response to Stress Meets Dedifferentiation'. *Planta* 233 (3): 433–38. https://doi.org/10.1007/s00425-011-1366-3.

Guo, Changkui, Yunmin Xu, Min Shi, Yongmin Lai, Xi Wu, Huasen Wang, Zhujun Zhu, R. Scott Poethig, and Gang Wu. 2017. 'Repression of miR156 by miR159 Regulates the Timing of the Juvenile-to-Adult Transition in Arabidopsis'. *The Plant Cell* 29 (6): 1293–304. https://doi.org/10.1105/tpc.16.00975.

Gutierrez, Laurent, John D. Bussell, Daniel I. Păcurar, Josèli Schwambach, Monica Păcurar, and Catherine Bellini. 2009. 'Phenotypic Plasticity of Adventitious Rooting in *Arabidopsis* Is Controlled by Complex Regulation of AUXIN RESPONSE FACTOR Transcripts and MicroRNA Abundance'. *The Plant Cell* 21 (10): 3119–32. https://doi.org/10.1105/tpc.108.064758.

Holec, Sarah, and Frédéric Berger. 2012. 'Polycomb Group Complexes Mediate Developmental Transitions in Plants'. *Plant Physiology* 158 (1): 35–43. https://doi.org/10.1104/pp.111.186445.

Horstman, Anneke, Mengfan Li, Iris Heidmann, Mieke Weemen, Baojian Chen, Jose M. Muino, Gerco C. Angenent, and Kim Boutilier. 2017. 'The BABY BOOM Transcription Factor ActivatestheLEC1-ABI3-FUS3-LEC2NetworktoInduceSomaticEmbryogenesis1[OPEN]'. *Plant Physiology* 175 (2): 848–57. https://doi.org/10.1104/pp.17.00232.

Ikeuchi, Momoko, Akira Iwase, Bart Rymen, Hirofumi Harashima, Michitaro Shibata, Mariko Ohnuma, Christian Breuer, et al. 2015. 'PRC2 Represses Dedifferentiation of Mature Somatic Cells in Arabidopsis'. *Nature Plants* 1 (7): 1–7. https://doi.org/10.1038/nplants.2015.89.

Ikeuchi, Momoko, Akira Iwase, Bart Rymen, Alice Lambolez, Mikiko Kojima, Yumiko Takebayashi, Jefri Heyman, et al. 2017. 'Wounding Triggers Callus Formation via Dynamic Hormonal and Transcriptional Changes'. *Plant Physiology* 175 (3): 1158–74. https://doi.org/10.1104/pp.17.01035.

Ikeuchi, Momoko, Akira Iwase, and Keiko Sugimoto. 2015. 'Control of Plant Cell Differentiation by Histone Modification and DNA Methylation'. *Current Opinion in Plant Biology*, Cell biology, 28 (December): 60–67. https://doi.org/10.1016/j.pbi.2015.09.004.

Ikeuchi, Momoko, Keiko Sugimoto, and Akira Iwase. 2013. 'Plant Callus: Mechanisms of Induction and Repression'. *The Plant Cell* 25 (9): 3159–73. https://doi.org/10.1105/tpc.113.116053.

Jha, Priyanka, and Vijay Kumar. 2018. 'BABY BOOM (BBM): A Candidate Transcription Factor Gene in Plant Biotechnology'. *Biotechnology Letters* 40 (11–12): 1467–75. https://doi.org/10.1007/s10529-018-2613-5.

Joshi, Sanjay, Priyanka Paul, Jeanne M. Hartman, and Sharyn E. Perry. 2022. 'AGL15 Promotion of Somatic Embryogenesis: Role and Molecular Mechanism'. *Frontiers in Plant Science* 13 (March): 861556. https://doi.org/10.3389/fpls.2022.861556.

Kakimoto, Tatsuo. 2001. 'Identification of Plant Cytokinin Biosynthetic Enzymes as Dimethylallyl Diphosphate: ATP/ADP Isopentenyltransferases'. *Plant and Cell Physiology* 42 (7): 677–85. https://doi.org/10.1093/pcp/pce112.

Karlova, Rumyana, Sjef Boeren, Eugenia Russinova, José Aker, Jacques Vervoort, and Sacco de Vries. 2006. 'The *Arabidopsis* SOMATIC EMBRYOGENESIS RECEPTOR-LIKE KINASE1 Protein Complex Includes BRASSINOSTEROID-INSENSITIVE1'. *The Plant Cell* 18 (3): 626–38. https://doi.org/10.1105/tpc.105.039412.

Kirschner, Gwendolyn K., Yvonne Stahl, Jafargholi Imani, Maria von Korff, and Rüdiger Simon. 2018. 'Fluorescent Reporter Lines for Auxin and Cytokinin Signalling in Barley (*Hordeum Vulgare*)'. *PLoS One* 13 (4): e0196086. https://doi.org/10.1371/journal.pone.0196086.

Kurup, Smita, John Runions, Uwe Köhler, Laurent Laplaze, Sarah Hodge, and Jim Haseloff. 2005. 'Marking Cell Lineages in Living Tissues'. *The Plant Journal* 42 (3): 444–53. https://doi.org/10.1111/j.1365-313X.2005.02386.x.

Laplaze, Laurent, Eva Benkova, Ilda Casimiro, Lies Maes, Steffen Vanneste, Ranjan Swarup, Dolf Weijers, et al. 2007. 'Cytokinins Act Directly on Lateral Root Founder Cells to Inhibit Root Initiation'. *The Plant Cell* 19 (12): 3889–3900. https://doi.org/10.1105/tpc.107.055863.

Lardon, Robin, Hoang Khai Trinh, Xiangyu Xu, Lam Dai Vu, Brigitte Van De Cotte, Markéta Pernisová, Steffen Vanneste, Ive De Smet, and Danny Geelen. 2022. 'Histidine Kinase Inhibitors Impair Shoot Regeneration in *Arabidopsis Thaliana via* Cytokinin Signaling and SAM Patterning Determinants'. *Frontiers in Plant Science* 13 (September): 894208. https://doi.org/10.3389/fpls.2022.894208.

Lardon, Robin, Erik Wijnker, Joost Keurentjes, and Danny Geelen. 2020. 'The Genetic Framework of Shoot Regeneration in *Arabidopsis* Comprises Master Regulators and Conditional Fine-Tuning Factors'. *Communications Biology* 3 (1): 1–13. https://doi.org/10.1038/s42003-020-01274-9.

Lee, Kyounghee, Jong Hee Kim, Ok-Sun Park, Yu Jin Jung, and Pil Joon Seo. 2022. 'Ectopic Expression of *WOX5* Promotes Cytokinin Signaling and de Novo Shoot Regeneration'. *Plant Cell Reports* 41 (12): 2415–22. https://doi.org/10.1007/s00299-022-02932-4.

Li, Hui, Mercedes Soriano, Jan Cordewener, Jose M. Muiño, Tjitske Riksen, Hiroyuki Fukuoka, Gerco C. Angenent, and Kim Boutilier. 2014. 'The Histone Deacetylase Inhibitor Trichostatin A Promotes Totipotency in the Male Gametophyte'. *The Plant Cell* 26 (1): 195–209. https://doi.org/10.1105/tpc.113.116491.

Li, Wei, Hui Liu, Zhi Juan Cheng, Ying Hua Su, Hua Nan Han, Yan Zhang, and Xian Sheng Zhang. 2011. 'DNA Methylation and Histone Modifications Regulate *De Novo* Shoot Regeneration in *Arabidopsis* by Modulating *WUSCHEL* Expression and Auxin Signaling'. *PLOS Genetics* 7 (8): e1002243. https://doi.org/10.1371/journal.pgen.1002243.

Liao, Che-Yang, Wouter Smet, Geraldine Brunoud, Saiko Yoshida, Teva Vernoux, and Dolf Weijers. 2015. 'Reporters for Sensitive and Quantitative Measurement of Auxin Response'. *Nature Methods* 12 (3): 207–10. https://doi.org/10.1038/nmeth.3279.

Liu, Boling, Ge Sun, Changju Liu, and Shijuan Liu. 2021. 'LEAFY COTYLEDON 2: A Regulatory Factor of Plant Growth and Seed Development'. *Genes* 12 (12): 1896. https://doi.org/10.3390/genes12121896.

Liu, Zhenhua, Juan Li, Long Wang, Qiang Li, Qing Lu, Yanchong Yu, Shuo Li, Ming-Yi Bai, Yuxin Hu, and Fengning Xiang. 2016. 'Repression of Callus Initiation by the miRNA-Directed Interaction of Auxin-Cytokinin in *Arabidopsis Thaliana*'. *The Plant Journal: For Cell and Molecular Biology* 87 (4): 391–402. https://doi.org/10.1111/tpj.13211.

Long, Jian-Mei, Chao-Yang Liu, Meng-Qi Feng, Yun Liu, Xiao-Meng Wu, and Wen-Wu Guo. 2018. 'miR156-*SPL* Modules Regulate Induction of Somatic Embryogenesis in Citrus Callus'. *Journal of Experimental Botany* 69 (12): 2979–93. https://doi.org/10.1093/jxb/ery132.

Matosevich, Rotem, and Idan Efroni. 2021. 'The Quiescent Center and Root Regeneration'. *Journal of Experimental Botany* 72 (19): 6739. https://doi.org/10.1093/jxb/erab319.

Müller-Xing, Ralf, and Qian Xing. 2022. 'The Plant Stem-Cell Niche and Pluripotency: 15 Years of an Epigenetic Perspective'. *Frontiers in Plant Science* 13 (October):1018559. https://doi.org/10.3389/fpls.2022.1018559.

Nameth, Blair, Steven J. Dinka, Steven P. Chatfield, Adam Morris, Jenny English, Dorrett Lewis, Rosalinda Oro, and Manish N. Raizada. 2013. 'The Shoot Regeneration Capacity of Excised Arabidopsis Cotyledons Is Established during the Initial Hours after Injury and Is Modulated by a Complex Genetic Network of Light Signalling'. *Plant, Cell & Environment* 36 (1): 68–86. https://doi.org/10.1111/j.1365-3040.2012.02554.x.

Pandey, Dhananjay K., and Bhupendra Chaudhary. 2014. 'Oxidative Stress Responsive SERK1 Gene Directs the Progression of Somatic Embryogenesis in Cotton (*Gossypium Hirsutum* L. Cv. Coker 310)'. *American Journal of Plant Sciences* 5 (1): 80–102. https://doi.org/10.4236/ajps.2014.51012.

Pandey, Dhananjay K., and Bhupendra Chaudhary. 2015. 'Genes and *Trans*-Factors Underlying Embryogenic Transition in Plant Soma-Cells'. In *Advances in the Understanding of Biological Sciences Using Next Generation Sequencing (NGS) Approaches*, edited by Gaurav Sablok, Sunil Kumar, Saneyoshi Ueno, Jimmy Kuo, and Claudio Varotto, 155–78. Cham: Springer International Publishing. https://doi.org/10.1007/978-3-319-17157-9_10.

Pandey, Dhananjay K., and Bhupendra Chaudhary. 2021. 'Transcriptional Loss of Domestication-Driven Cytoskeletal *GhPRF1* Gene Causes Defective Floral and Fiber Development in Cotton (*Gossypium*)'. *Plant Molecular Biology* 107 (6): 519–32. https://doi.org/10.1007/s11103-021-01200-5.

Pandey, Dhananjay K., and Bhupendra Chaudhary. 2024. 'Ectopic Overexpression of Domestication-Driven Cytoskeletal *Profilin* (*GhPRF1*) Gene Improves Flowering and Trichome Development in Cotton (*Gossypium Hirsutum* L.)'. *Journal of Plant Growth Regulation*, May. https://doi.org/10.1007/s00344-024-11329-4.

Pandey, Dhananjay Kumar, and Bhupendra Chaudhary. 2014. 'Role of Plant Somatic Embryogenesis Receptor Kinases (SERKs) in Cell-to-Embryo Transitional Activity: Key at Novel Assorted Structural Subunits'. *American Journal of Plant Sciences* 5 (21): 3177–93. https://doi.org/10.4236/ajps.2014.521334.

Pernisova, Marketa, Martina Grochova, Tomas Konecny, Lenka Plackova, Danka Harustiakova, Tatsuo Kakimoto, Marcus G. Heisler, Ondrej Novak, and Jan Hejatko. 2018. 'Cytokinin Signalling Regulates Organ Identity via AHK4 Receptor in *Arabidopsis*'. *Development* 145: dev163907. https://doi.org/10.1242/dev.163907.

Pernisová, Markéta, Petr Klíma, Jakub Horák, Martina Válková, Jiří Malbeck, Přemysl Souček, Pavel Reichman, et al. 2009. 'Cytokinins Modulate Auxin-Induced Organogenesis in Plants via Regulation of the Auxin Efflux'. *Proceedings of the National Academy of Sciences of the United States of America* 106 (9): 3609–14. https://doi.org/10.1073/pnas.0811539106.

Qiao, Meng, Zhongjuan Zhao, Yuguang Song, Zhenhua Liu, Lingxue Cao, Yanchong Yu, Shuo Li, and Fengning Xiang. 2012. 'Proper Regeneration from *In Vitro* Cultured *Arabidopsis Thaliana* Requires the microRNA-Directed Action of an Auxin Response Factor'. *The Plant Journal: For Cell and Molecular Biology* 71 (1): 14–22. https://doi.org/10.1111/j.1365-313X.2012.04944.x.

Reid, James B., and John J. Ross. 2011. 'Regulation of Tissue Repair in Plants'. *Proceedings of the National Academy of Sciences of the United States of America* 108 (42): 17241–42. https://doi.org/10.1073/pnas.1114432108.

Reuveni, Moshe, and Dalia Evenor. 2007. 'On the Effect of Light on Shoot Regeneration in Petunia'. *Plant Cell, Tissue and Organ Culture* 89 (1): 49–54. https://doi.org/10.1007/s11240-007-9215-6.

Riou-Khamlichi, Catherine, Rachael Huntley, Annie Jacqmard, and James A.H. Murray. 1999. 'Cytokinin Activation of Arabidopsis Cell Division through a D-Type Cyclin'. *Science* 283 (5407): 1541–44. https://doi.org/10.1126/science.283.5407.1541.

Savina, Maria S., Taras Pasternak, Nadya A. Omelyanchuk, Klaus Palme, and Viktoriya V. Lavrekha. 2020. 'Cell Dynamics in WOX5-Overexpressing Root Tips: The Impact of Local Auxin Biosynthesis'. *Frontiers in Plant Science* 11 (October). https://doi.org/10.3389/fpls.2020.560169.

Schepetilnikov, Mikhail, and Lyubov A. Ryabova. 2017. 'Auxin Signaling in Regulation of Plant Translation Reinitiation'. *Frontiers in Plant Science* 8 (June). https://doi.org/10.3389/fpls.2017.01014.

Sena, Giovanni, and Kenneth D. Birnbaum. 2010. 'Built to Rebuild: In Search of Organizing Principles in Plant Regeneration'. *Current Opinion in Genetics & Development* 20 (4): 460–65. https://doi.org/10.1016/j.gde.2010.04.011.

Shen, Chia-Ning, Zoë D Burke, and David Tosh. 2004. 'Transdifferentiation, Metaplasia and Tissue Regeneration'. *Organogenesis* 1 (2): 36–44. https://doi.org/10.4161/org.1.2.1409.

Shim, Sangrea, Hee Kyoung Kim, Soon Hyung Bae, Hoonyoung Lee, Hyo Ju Lee, Yu Jin Jung, and Pil Joon Seo. 2020. 'Transcriptome Comparison between Pluripotent and Non-Pluripotent Calli Derived from Mature Rice Seeds'. *Scientific Reports* 10 (December):21257. https://doi.org/10.1038/s41598-020-78324-z.

Skoog, F., and C. O. Miller. 1957. 'Chemical Regulation of Growth and Organ Formation in Plant Tissues Cultured in Vitro'. *Symposia of the Society for Experimental Biology* 11: 118–30.

Steward, F. C., P. V. Ammirato, and M. O. Mapes. 1970. 'Growth and Development of Totipotent Cells: Some Problems, Procedures, and Perspectives'. *Annals of Botany* 34 (4): 761–87. https://doi.org/10.1093/oxfordjournals.aob.a084408.

Strotmann, Vivien I., and Yvonne Stahl. 2021. 'At the Root of Quiescence: Function and Regulation of the Quiescent Center'. *Journal of Experimental Botany* 72 (19): 6716–26. https://doi.org/10.1093/jxb/erab275.

Su, Ying H., Xiang Y. Zhao, Yu B. Liu, Chuan L. Zhang, Sharman D. O'Neill, and Xian S. Zhang. 2009. 'Auxin-Induced *WUS* Expression Is Essential for Embryonic Stem Cell Renewal during Somatic Embryogenesis in Arabidopsis'. *The Plant Journal* 59 (3): 448–60. https://doi.org/10.1111/j.1365-313X.2009.03880.x.

Su, Ying Hua, Zhi Juan Cheng, Yu Xiao Su, and Xian Sheng Zhang. 2010. 'Pattern Analysis of Stem Cell Differentiation during in Vitro Arabidopsis Organogenesis'. *Frontiers in Biology* 5 (5): 464–70. https://doi.org/10.1007/s11515-010-0820-0.

Su, Ying Hua, and Xian Sheng Zhang. 2014. 'The Hormonal Control of Regeneration in Plants'. In *Current Topics in Developmental Biology*, edited by Brigitte Galliot, 108: 35–69. Mechanisms of Regeneration. Academic Press. https://doi.org/10.1016/B978-0-12-391498-9.00010-3.

Sugimoto, Kaoru, Sean P. Gordon, and Elliot M. Meyerowitz. 2011. 'Regeneration in Plants and Animals: Dedifferentiation, Transdifferentiation, or Just Differentiation?' *Trends in Cell Biology* 21 (4): 212–18. https://doi.org/10.1016/j.tcb.2010.12.004.

Sugimoto, Kaoru, Yuling Jiao, and Elliot M. Meyerowitz. 2010. '*Arabidopsis* Regeneration from Multiple Tissues Occurs via a Root Development Pathway'. *Developmental Cell* 18 (3): 463–71. https://doi.org/10.1016/j.devcel.2010.02.004.

Sun, Tai-Ping. 2008. 'Gibberellin Metabolism, Perception and Signaling Pathways in Arabidopsis'. *The Arabidopsis Book/American Society of Plant Biologists* 6 (September): e0103. https://doi.org/10.1199/tab.0103.

Sussex, Ian M. 2008. 'The Scientific Roots of Modern Plant Biotechnology'. *The Plant Cell* 20 (5): 1189–98. https://doi.org/10.1105/tpc.108.058735.

Suzuki, Masaharu, Heidi H.-Y. Wang, and Donald R. McCarty. 2007. 'Repression of the LEAFY COTYLEDON 1/B3 Regulatory Network in Plant Embryo Development by VP1/ABSCISIC ACID INSENSITIVE 3-LIKE B3 Genes'. *Plant Physiology* 143 (2): 902–11. https://doi.org/10.1104/pp.106.092320.

Tanaka, Motoki, Akira Kikuchi, and Hiroshi Kamada. 2008. 'The Arabidopsis Histone Deacetylases HDA6 and HDA19 Contribute to the Repression of Embryonic Properties after Germination'. *Plant Physiology* 146 (1): 149–61. https://doi.org/10.1104/pp.107.111674.

Thorpe, Trevor. 2012. 'History of Plant Tissue Culture'. In *Plant Cell Culture Protocols*, edited by Víctor M. Loyola-Vargas and Neftalí Ochoa-Alejo, 877: 9–27. Methods in Molecular Biology. Totowa, NJ: Humana Press. https://doi.org/10.1007/978-1-61779-818-4_2.

Thorpe, Trevor A. 2007. 'History of Plant Tissue Culture'. *Molecular Biotechnology* 37 (2): 169–80. https://doi.org/10.1007/s12033-007-0031-3.

Wang, Long, Zhenhua Liu, Meng Qiao, and Fengning Xiang. 2018. '*miR393* Inhibits *In Vitro* Shoot Regeneration in *Arabidopsis Thaliana* via Repressing *TIR1*'. *Plant Science* 266 (January): 1–8. https://doi.org/10.1016/j.plantsci.2017.10.009.

Wang, Xingchun, Qi-Wen Niu, Chong Teng, Chao Li, Jinye Mu, Nam-Hai Chua, and Jianru Zuo. 2009. 'Overexpression of PGA37/MYB118 and MYB115 Promotes Vegetative-to-Embryonic Transition in Arabidopsis'. *Cell Research* 19 (2): 224–35. https://doi.org/10.1038/cr.2008.276.

Wójcikowska, Barbara, Karolina Jaskóła, Przemysław Gąsiorek, Magdalena Meus, Katarzyna Nowak, and Małgorzata D. Gaj. 2013. '*LEAFY COTYLEDON2* (*LEC2*) Promotes Embryogenic Induction in Somatic Tissues of Arabidopsis, via *YUCCA*-Mediated Auxin Biosynthesis'. *Planta* 238 (3): 425–40. https://doi.org/10.1007/s00425-013-1892-2.

Xu, Jian, Hugo Hofhuis, Renze Heidstra, Michael Sauer, Jirí Friml, and Ben Scheres. 2006. 'A Molecular Framework for Plant Regeneration'. *Science* 311 (5759): 385–88. https://doi.org/10.1126/science.1121790.

Xu, Mingli, Tieqiang Hu, Jianfei Zhao, Mee-Yeon Park, Keith W. Earley, Gang Wu, Li Yang, and R. Scott Poethig. 2016. 'Developmental Functions of miR156-Regulated *SQUAMOSA PROMOTER BINDING PROTEIN-LIKE* (*SPL*) Genes in *Arabidopsis Thaliana*'. *PLOS Genetics* 12 (8): e1006263. https://doi.org/10.1371/journal.pgen.1006263.

Yang, Woorim, Myung-Hwan Choi, Bosl Noh, and Yoo-Sun Noh. 2020. 'De Novo Shoot Regeneration Controlled by HEN1 and TCP3/4 in Arabidopsis'. *Plant and Cell Physiology* 61 (9): 1600–1613. https://doi.org/10.1093/pcp/pcaa083.

Zhang, Tian-Qi, Heng Lian, Hongbo Tang, Karel Dolezal, Chuan-Miao Zhou, Sha Yu, Juan-Hua Chen, et al. 2015. 'An Intrinsic MicroRNA Timer Regulates Progressive Decline in Shoot Regenerative Capacity in Plants'. *The Plant Cell* 27 (2): 349–60. https://doi.org/10.1105/tpc.114.135186.

Zhang, Ye, Masaaki Umeda, and Tatsuo Kakimoto. 2022. 'Pericycle Cell Division Competence Underlies Various Developmental Programs'. *Plant Biotechnology* 39 (1): 29–36. https://doi.org/10.5511/plantbiotechnology.21.1202a.

Zheng, Yumei, Na Ren, Huai Wang, Arnold J. Stromberg, and Sharyn E. Perry. 2009. 'Global Identification of Targets of the Arabidopsis MADS Domain Protein AGAMOUS-Like15'. *The Plant Cell* 21 (9): 2563–77. https://doi.org/10.1105/tpc.109.068890.

3 Exploring the Role of Plant Growth Regulators in Regeneration

Deepshikha Thakur, Dhananjay Suresh Gaikwad, Abha Sharma, and Sky

3.1 INTRODUCTION

Plant regeneration is the major outcome of plant tissue culture where somatic embryogenesis and organogenesis are frequently experimented for the regeneration of genetically transformed plants. The process by which somatic cells develop into plants is called somatic or asexual embryogenesis while the process of organ formation under the influence of several hormones *in vitro* is termed organogenesis. Somatic embryogenesis allows the utilization of somatic embryos as synthetic seeds; however, application of somatic embryogenesis is not restricted to synthetic seed technology. Somatic embryogenesis also forms a major part of the plant regeneration system where hybridization, ploidy evaluation of somatic cells, virus elimination, metabolite production, and *in vitro* mycorrhizal initiation have been investigated. Hormones play a vital role in the regeneration of tissues and organs, serving as key regulators of growth, repair, and cellular function.

3.2 PLANT GROWTH REGULATORS

The regulation of plant growth and development is intricately governed by a diverse array of phytohormones, each playing unique yet interconnected roles in shaping various physiological processes. Among these essential plant growth regulators, auxins and gibberellins take the lead, orchestrating cell elongation, division, and differentiation, thereby influencing overall plant architecture and development (Figure 3.1). Cytokinins, on the other hand, are key regulators of cell division and differentiation, promoting growth in meristematic tissues and delaying senescence. Abscisic acid, known for its role in stress responses, regulates stomatal closure and seed dormancy, ensuring plant survival under adverse environmental conditions. Ethylene, a gaseous hormone, regulates diverse processes such as fruit ripening, leaf senescence, and response to mechanical stress. Brassinosteroids, a group of steroid hormones, play crucial roles in cell expansion, vascular differentiation, and stress responses. Lastly, jasmonic acid emerges as a versatile regulator, mediating defense responses against biotic stresses, modulating reproductive development, and integrating various hormonal signals for optimal growth and adaptation. Together, these plant growth

DOI: 10.1201/9781032702391-3

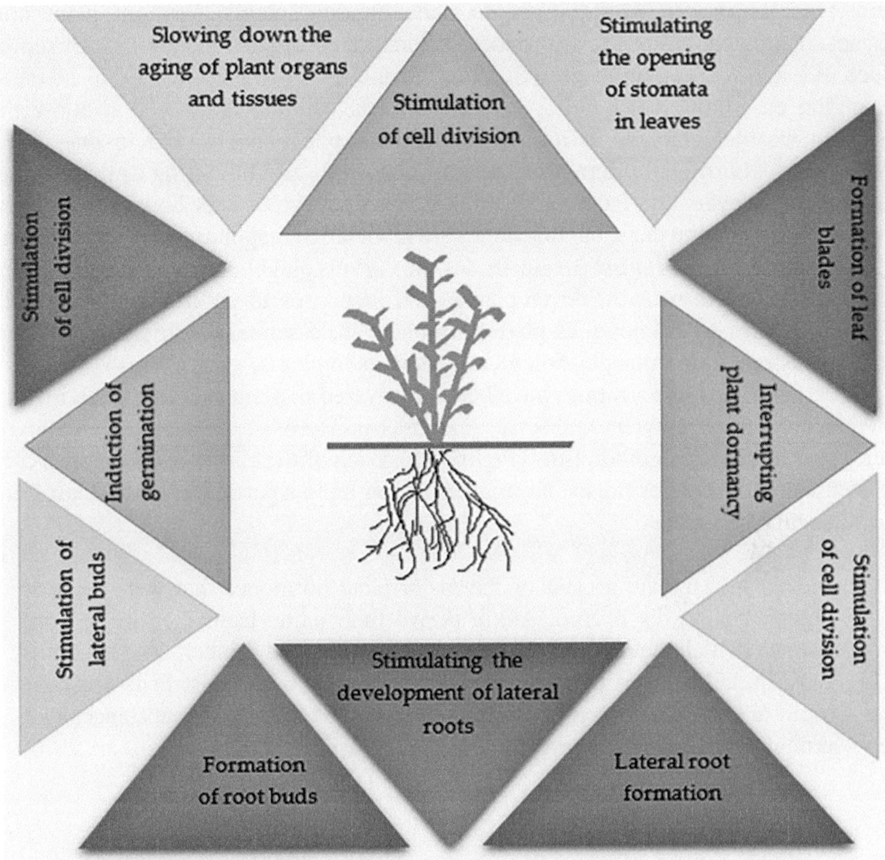

FIGURE 3.1 Roles of phytohormones. Auxins (depicted in dark grey), cytokinins (shown in light grey), and gibberellins (represented in mid grey). (Sosnowski et al., 2023.)

regulators form a complex network that finely tunes plant growth and development in response to internal and external cues, ensuring the resilience and adaptation of plants in diverse environments (Coggins and Lovatt, 2014).

3.3 TYPES OF PLANT GROWTH REGULATORS IN REGENERATION

Plant growth regulators (PGRs), sometimes referred to as phytohormones or plant hormones, are crucial for controlling a wide range of physiological processes related to plant growth and development (Naqvi, 2001). Regeneration, which involves tissue culture and propagation techniques, and hormones found in plants are especially significant (Soliman, 2019). Plant hormones control the complete plant lifecycle, including germination, rooting, growth, flowering, fruit, foliage, and death (Taiz et al., 2015). Plant hormones are natural chemicals present in tiny amounts inside plants. They serve as an internal chemical signaling system that dictates what the plant should do at a particular time (Miransari and Smith, 2014). Understanding how plant hormones work is key to understanding plants. Without plant hormones, plants would not grow, produce differentiated organs, metabolize, or reproduce. As well as regulating growth, plant hormones tell the plant how to adapt to changing environmental conditions. They induce cellular changes that allow the plant to weather changes in its surroundings despite being rooted in one place. For example, hormonal interactions control leaf drops and browning as the weather gets colder (Weyers and Paterson, 2001a). It is not known exactly how the plant perceives environmental cues, but hormones are involved in responding to them (Weyers and Paterson, 2001b). Figure 3.1 illustrates the various plant hormones and their roles in regulating plant growth, development, and responses to environmental stimuli. Plant hormones, also known as phytohormones, are essential chemical messengers that orchestrate numerous physiological processes in plants.

Specific groups of hormones have been discovered in plants so far, which include auxins, cytokinins, ethylenes, abscisic acid, brassinosteroids, strigolactones, salicylic acid, jasmonates, and gibberellins (Figure 3.2). Each of the classes of hormones contains many different hormones. Hormones belonging to a particular class share similar molecular structures.

1. *Auxins*: Auxins are a class of the major plant hormones that were the first phytohormone discovered. Auxin derive their name from the Greek word '*auxein*' which means "to grow/increase" (Enders and Strader, 2015). Auxins were initially linked to plant growth by the Dutch biologist Frits Warmolt Went in the 1920s. Auxin is a general term used to indicate substances that promote root and shoot development in plant tissues. (Overvoorde et al., 2010)

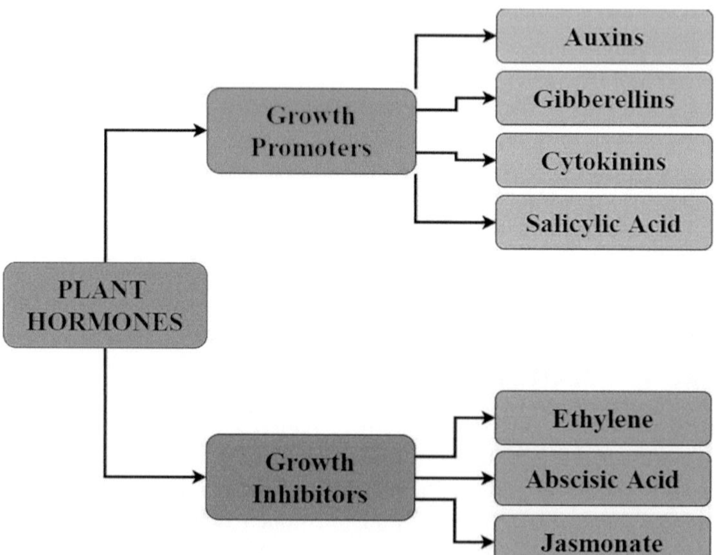

FIGURE 3.2 Classification of Phytohormones: A Detailed Overview of Key Hormones (Asif et al., 2022.)

They stimulate cell elongation and are often used in tissue culture to induce root formation in cuttings or callus cultures. The most potent native auxin and produces the bulk of auxin actions in intact plants, indole 3-acetic acid (IAA) is the most significant member of the auxin family (Ulmasov et al., 1997). Natural auxins include indole-3-acetonitrile (IAN) and phenylacetic acid, while synthetic auxins comprise indole-3-butyric acid (IBA), alpha-naphthyl acetic acid (NAA), and 1-naphthyl acetamide (NAD). Auxin, a key plant hormone, exerts a multitude of functions crucial for plant growth and development. Firstly, it plays a pivotal role in stimulating internode elongation, facilitating the upward growth of stems. Additionally, auxin promotes leaf growth, aiding in the expansion and development of foliage. Moreover, it serves as a stimulant for the initiation of vascular tissue, crucial for the transport of water, nutrients, and hormones throughout the plant, as well as for fruit growth, influencing their size and development. Interestingly, auxin also exhibits inhibitory effects on root growth, regulating the balance between shoot and root development. However, it paradoxically stimulates the initiation of roots on stem cuttings, highlighting its dual role in root development. Furthermore, auxin plays a significant role in the differentiation of vascular tissue, promoting the formation of both xylem and phloem, essential components of the plant's transport system. In tissue culture, auxin facilitates lateral root development through adventitious rooting, enhancing the ability of plants to establish themselves in new environments. Moreover, auxin mediates tropic responses, such as bending toward gravity (gravitropism) and light (phototropism), enabling plants to adapt and optimize their growth in response to environmental cues. In summary, auxin's multifaceted functions encompassing stem elongation, leaf and fruit growth, vascular tissue differentiation, root development, and tropic responses underscore its indispensable role in orchestrating various aspects of plant growth and development.

3.4 MECHANISM OF ACTION OF AUXIN

The apical meristem of shoots, immature leaves, and seeds in where auxin is mostly produced. Auxin is unidirectional, polar substance that travels downwards from its synthesis site. Auxin concentration gradient, produced by polar transport cause particular reactions. Transport of auxin out of the cell is regulated by auxin-specific transport proteins in the plasma membrane. Plant hormones function by signal transduction, which triggers several cellular reactions by binding to receptors connected to enzymes, and auxin facilitates the catalysis of a process. Auxin causes a repressor protein for some genes (auxin response gene) to bind to ubiquitin when it interacts with the receptor. This leads to the repressor protein's destruction and the progression of auxin response gene transcription, which promotes cellular growth and development (Woodward and Bartel, 2005). Auxin, a crucial phytohormone, exerts a profound influence on various aspects of plant growth and development, as extensively documented in recent research (Li et al., 2023). Its involvement in parthenocarpy, the production of seedless fruits, underscores its significance in fruit development, where precise hormonal regulation is essential for optimal yield and quality. Auxin governs callus formation and morphogenesis, crucial for wound healing and regeneration in plants. Its role in secondary growth ensures the development of sturdy stems and branches, contributing to plant architecture and structural support. Furthermore, auxin promotes root initiation on cuttings, facilitating vegetative propagation and enhancing the efficiency of horticultural practices. The concept of apical dominance, where the terminal bud suppresses the growth of lateral buds, is intricately tied to auxin signaling, highlighting its importance in shaping plant architecture and optimizing resource allocation. Auxin's involvement in flowering regulation underscores its role in coordinating developmental transitions critical for reproductive success. Auxin's interaction with herbicides underscores its significance in agricultural practices, where its modulation can influence weed control strategies and crop productivity. Finally, auxin promotes tropism, guiding plant organs toward or away from environmental stimuli such as light or gravity, showcasing its adaptive significance in plant survival and growth (Li et al., 2023). Auxin emerges as a master regulator of plant growth and development, with its multifaceted roles extending from fundamental cellular processes to complex morphogenetic events, profoundly shaping the biology and ecology of plants (Li et al., 2023). Understanding its mechanisms of action holds immense potential for agricultural innovation and ecosystem management.

3.4.1 Gibberellins (GAs)

Gibberellins are plant hormones that regulate various developmental processes, including stem elongation, germination, dormancy, flowering, flower development, and leaf and fruit senescence (Hedden and Sponsel, 2015). One of the most well-known gibberellins is GA3, which has been extensively studied for its effects on plant growth and development. However, researchers have identified a total of around 50 different gibberellins from various sources, including bacteria, fungi, and plants. Among these, approximately 40 are found in green plants. The unintentional breeding of crop strains lacking in gibberellin synthesis could lead to stunted growth, reduced yield, and altered developmental patterns in plants. Therefore, understanding the role of gibberellins and their synthesis pathways is essential for optimizing crop production and agricultural practices.

3.5 FUNCTIONS OF GIBBERELLINS

Gibberellin hormones serve a multitude of functions in plant physiology, influencing various growth and developmental processes. For instance, they stimulate the abrupt extension of internodes just before blooming in rosette plants like cabbage and beets, a phenomenon known as bolting. Additionally, gibberellins postpone senescence, the aging process in plants, thereby extending the productive lifespan of crops. Gibberellins induce parthenocarpy, the development of seedless fruits, and have the remarkable ability to reverse dwarfism in plants by promoting stem elongation. They can also influence sexual differentiation, triggering certain plants to develop male characteristics, as observed in cannabis. In terms of seed germination, gibberellins play a crucial role by breaking dormancy and promoting the synthesis of hydrolytic enzymes such as lipase and amylase in the endosperm of sprouting cereal grains and barley seeds. In agriculture, gibberellins find widespread applications (Figure 3.1). They are used to elongate genetically dwarf plants, induce bolting and flowering in long-day plants, and substitute for cold treatment to break dormancy, thereby promoting uniform germination. Gibberellins are also employed to induce parthenocarpy in fruit crops, enhancing fruit yield and quality. Gibberellins have been tested in leaf and root cultures of digitalis, leading to higher production of digoxin, a valuable pharmaceutical compound. Additionally, spraying juvenile conifers with gibberellins accelerates their maturity period, facilitating early seed production. Similarly, applying gibberellins to sugarcane crops increases stem length, which can enhance yield and facilitate harvesting. Overall, the versatile functions and applications of gibberellins underscore their significance in agriculture and plant biotechnology (Hedden and Sponsel, 2015). Gibberellins (GAs) are crucial plant hormones that significantly influence plant regeneration by promoting cell elongation, differentiation, and tissue development. In plant tissue culture, GAs enhance the regeneration process by stimulating the growth of shoots and roots from explants, facilitating the formation of new plant structures. They work synergistically with other hormones, such as cytokinins and auxins, to regulate key developmental processes and improve the efficiency of tissue regeneration. GAs are particularly effective in overcoming dormancy and triggering seed germination, which is essential for successful plant propagation and recovery from tissue damage (Wang et al., 2011; Olszewski et al., 2002).

3.6 CYTOKININS

Plant hormones known as cytokinins (CK) stimulate cell division in both the roots and shoots of plants. Their main functions are in cell differentiation and growth. They have an impact on leaf senescence, axillary bud development, and apical dominance. (Oppenheim, 2001). Cytokinins are a class of compounds with a chemical structure resembling adenine. They are widely distributed in nature, being found in almost all higher plants, as well as in mosses, fungi, bacteria, and various other organisms. With over 200 natural and synthetic cytokinins identified to date, these compounds play diverse roles in plant growth and development. The discovery of cytokinins dates back to 1961 when the first naturally occurring cytokinin, zeatin, was isolated from corn by Miller. Alongside zeatin, other naturally occurring cytokinins include N6 dimethyl amino purine and isopentenyl aminopurine. Additionally, synthetic cytokinins such as kineatin, adenine, 6-benzyl adenine benzimidazole, and N,N-diphenyl urea have been developed. Cytokinins exert various functions in plant physiology. One of their primary roles is to stimulate cell division, earning them the name "cytokines." They also play a crucial role in morphogenesis, promoting shoot initiation and bud formation, particularly in tissue culture techniques like micropropagation. Moreover, cytokinins contribute to the growth of lateral roots, aiding in the establishment of a well-developed root system. Overall, cytokinins are essential regulators of plant growth and development, influencing processes ranging from cell division to organogenesis. Their diverse functions make them valuable tools in agricultural practices, biotechnology, and research aimed at understanding plant biology and enhancing crop productivity (Márquez-López et al., 2019). Cytokinins are pivotal in plant regeneration as they stimulate cell division and differentiation, which are crucial for the formation of new tissues from callus cultures. These hormones work in concert with auxins to drive organogenesis, influencing whether regenerated tissues form shoots or roots depending on their concentration and balance. Cytokinins also play a role in stress response, helping plants recover from damage and adapt to adverse conditions. Their regulatory effects on cell division and developmental processes make them indispensable for successful plant tissue culture and regeneration (Gaj, 2004; Kieber and Schaller, 2014).

3.7 ABSCISIC ACID (ABA)

Abscisic acid is a plant hormone. ABA is involved in a variety of plant developmental processes, such as the regulation of organ size, stomatal closure, and seed and bud dormancy. When it comes to how plants react to environmental stressors including drought, salinity in the soil, freezing and cold tolerance, heat stress, and heavy metal ion tolerance, it is particularly crucial (Sah et al., 2016; Kumar et al., 2022). ABA is often considered a growth inhibitor, and its role is more related to stress responses and dormancy. In tissue culture, controlling ABA levels can be important for optimizing regeneration conditions.

3.8 ETHYLENE

Plant hormone ethylene affects the ripening and rotting of plants. This specific plant hormone is very interesting since it manifests as a gas (Pierik et al., 2007). There is no other plant hormone that is gaseous. Plants can produce ethylene in almost any part and can diffuse through its tissue outside the plant, and travel through the air to affect a completely different plant. Some of the functions of the Ethylene Hormones speeds up the ripening of fruits, Controls leaf epinasty, Breaks the dormancy of seeds and buds, Accelerates the elongation of petioles and internodes, encourages the withering and abscission of leaves and blossoms, promotes the growth of roots and the formation of root hair, increasing the absorption surface, In monoecious plants, it stimulates femaleness, Formation of apical hooks in dicot seedlings (Iqbal et al., 2017).

3.9 IMPORTANT FUNCTIONS OF ETHYLENE

Ethylene causes two processes in plants: reduction in stem elongation and acceleration of senescence. Besides, it also helps in ripening of fruits, initiating germination in peanut seeds, sprouting of potato tuber, promoting root growth and root-hair formation, inducing flowering in mango and also ethylene is the most widely used plant growth hormone in agriculture (Iqbal et al., 2017).

3.10 BRASSINOSTEROIDS

Plant hormones belonging to the class brassinosteroids are present in both higher plants and more basic species like algae and bryophytes. Today, there are roughly 60 recognized varieties with comparable molecular structures. Pollen contains the largest quantities of brassinosteroids in plants. In addition, they are present in the grains, anthers, seeds, leaves, stems, roots, flowers, and younger tissue, which often has higher concentrations than older tissue (Asami et al., 2005). The growth and development of plants depend on brassinosteroids, much like they do on other groups of plant hormones such as auxins, cytokinins, and gibberellins. Nonetheless, brassinosteroids function slightly differently and operate at far lower concentrations than other groups of plant hormones. Brassinolide is a well-known brassinosteroid (Fridman and Savaldi-Goldstein, 2013).

3.11 JASMONIC ACID

Hormones are used for signaling the networks that regulate stress responses have been identified. Plant hormones are natural and non-toxic compounds, which could be applied as safe and eco-friendly chemical control agents (Ruan et al., 2019). Jasmonic acid has a role in a number of stress responses and defense mechanisms in plants. It may influence regeneration processes, especially under stress conditions (Ali and Baek, 2020). The specific combination and concentration of these PGRs can be manipulated in tissue culture techniques to optimize regeneration protocols for different plant species. The balance between auxins and cytokinins, in particular, plays a crucial role in determining whether roots, shoots, or both will be induced in cultured tissues (Ravnikar et al., 1993).

3.12 MECHANISMS OF PLANT REGENERATION

Regeneration of plants occurs when plants replace or repair damaged structures based on the pluripotency and totipotency of their cells. One of the most popular regeneration technologies is tissue culture.

The process of growing an entire plant from an explant or even just one plant cell under aseptic conditions is called plant tissue culture. This aspect of plant biotechnology depends on the phenomenon known as cell totipotency, which describes any single cell's capacity to produce all the differentiated cells that are characteristic of organs and to regenerate into a full plant (Trigiano and Gray, 2016). The rapid mass multiplication of elite genotypes on huge sizes in a relatively short amount of time is made possible by micropropagation, which takes the use of this essential characteristic of plant cells. Micropropagation now plays a significant role in agriculture, horticulture, and industry thanks to the year-round generation of healthy seedlings and the shortening of the vegetative cycle (Suman, 2017). It is also a key method for genetic engineering, agricultural enhancement, and the propagation of new kinds resulting from somaclonal variation.

3.13 PLANT TISSUE CULTURE TECHNOLOGY

Micropropagation, an *in vitro* technique, enables the rapid multiplication of plant material from explants, revolutionizing plant production. The process of micropropagation consists of six distinct stages:

1. *Stage 0*: Pre-treatment and immobilization of plant stock.
2. *Stage I*: Establishment of initial cultures.
3. *Stage II*: Elongation and multiplication of shoots.
4. *Stage III*: Root induction.
5. *Stage IV*: Hardening and acclimatization of plants.
6. *Stage V*: Transfer of plants to natural field conditions.

The first four stages are carried out in a sterile environment to prevent contamination from external pathogens (Orlikowska et al., 2017).

3.14 SOMATIC EMBRYOGENESIS IN PLANT REGENERATION

Somatic embryogenesis highlights the totipotent nature of plant cells, where somatic cells dedifferentiate and form embryonic structures capable of developing into complete plants. This process can be initiated by stress, hormonal stimuli (notably auxins), or alterations in gene expression (Horstman et al., 2017). Somatic embryos can form directly from explants or indirectly via an intermediary callus stage (Yang and Zhang, 2010).

In indirect somatic embryogenesis, an embryogenic callus is formed, which eventually gives rise to proembryonic masses that develop into somatic embryos, ultimately differentiating into shoots and roots under specific culture conditions. To induce shoot formation, cytokinin-rich media are required, while auxin-dominant media are necessary for root induction (Fitch and Manshardt, 1990). This indirect process, although longer, can yield a greater number of regenerated plantlets due to the extensive callus growth (Gaj, 2011).

In direct somatic embryogenesis, embryos emerge directly from explant tissues, bypassing the callus stage, leading to faster regeneration (Zhang et al., 2021). While direct embryogenesis is quicker, the indirect method is advantageous when explant availability is limited or large-scale plant production is desired for agricultural applications.

3.15 MOLECULAR MECHANISMS OF SOMATIC EMBRYOGENESIS

Key molecular pathways regulate somatic embryogenesis, particularly the WUS-CLV3 and STM-CUC signaling networks, which control shoot stem cell homeostasis. The expression of the WUSCHEL (WUS) gene, which occurs shortly after induction on shoot-inducing media, is essential for shoot progenitor cell development (Zhang et al., 2021). Overexpression of WUS leads to excessive shoot formation, whereas WUS mutants lose the capacity to regenerate shoots entirely (Gordon et al., 2007). Auxins and cytokinins intricately control WUS expression. Type-B response regulators (ARR1, ARR2, ARR10, ARR12) enhance WUS transcription while concurrently inhibiting auxin synthesis through YUC regulation (Zhang et al., 2021). A negative feedback loop involving type-A ARRs (ARR5, ARR6, ARR7, ARR15) modulates cytokinin signaling, balancing the overall hormonal influence on shoot regeneration (Sugimoto et al., 2011).

3.16 ROLE OF PLANT GROWTH REGULATORS IN REGENERATION

The significance of plant hormones in tissue culture was first predicted by Haberlandt in 1902, envisioning that vegetative cells could eventually be used for embryo cultivation under proper *in vitro* conditions (Haberlandt, 1902). This vision materialized in subsequent decades, with breakthroughs in plant regeneration technologies.

Auxins, particularly indole-3-acetic acid (IAA), play a critical role in both direct and indirect regeneration processes. In indirect regeneration, de novo organs are produced from a callus, whereas in direct regeneration, organs are induced directly from explants. The pioneering work of Skoog and Miller (1957) demonstrated that the ratio of auxin to cytokinin dictates the type of organ formed. High auxin-to-cytokinin ratios favor root formation, while high cytokinin-to-auxin ratios induce shoot regeneration.

This discovery laid the foundation for understanding how hormonal balances govern the regeneration pathways in plant tissue culture. The appropriate combination of auxins and cytokinins determines not only whether callus development occurs but also whether this callus gives rise to roots, shoots, or other plant tissues.

3.17 HORMONAL REGULATION AND PLANT REGENERATION

Plant growth and development are governed by a diverse array of plant hormones that influence gene expression to regulate key physiological processes (Huang et al., 2015). One of the most fascinating attributes of plant cells is their totipotency, which enables them to regenerate complete organisms through embryogenesis (Horstman et al., 2017). The interactions among major plant growth regulators, including auxins, cytokinins, gibberellins, and abscisic acid, are critical in orchestrating the intricate processes of plant regeneration (Suge, 1984).

Auxins play a pivotal role in nearly every stage of plant growth and development, ranging from embryogenesis to senescence. They are particularly vital for initiating somatic embryogenesis and organogenesis. For instance, auxins facilitate cellular reprogramming, enabling differentiated cells to acquire embryogenic competence (Huang et al., 2015). Cytokinins, in contrast, primarily stimulate cell division and are essential for the development of shoot meristems. Gibberellins complement these functions by promoting cell elongation, contributing to the overall success of regeneration (Brian, 1959). The balance and interaction between these hormones significantly influence the outcome of regeneration processes.

In vitro plant tissue culture techniques have proven invaluable for studying the role of these growth regulators in plant regeneration. Liquid culture systems, in particular, provide an optimal environment with a consistent supply of water, nutrients, and growth regulators, thereby enhancing the propagation of shoots, roots, bulblets, and somatic embryos (Ascough and Fennell, 2004). Additionally, the controlled physical conditions in liquid cultures—such as reduced gradients and the dilution of inhibitory exudates—further support morphogenic development, including meristemoid formation and somatic embryogenesis (Ascough and Fennell, 2004).

Research has consistently highlighted the significance of precise auxin-to-cytokinin ratios in successful plant regeneration through somatic embryogenesis.

These hormones often act synergistically, inducing the expression of key transcription factors, such as the *BABY BOOM* (BBM) gene, which is a crucial regulator of somatic embryo formation (Horstman et al., 2017). The regulation of developmental processes such as somatic embryogenesis, storage organ development, and meristemoid formation involves dynamic and highly coordinated hormonal interactions, necessitating further molecular-level investigations (Ascough and Fennell, 2004).

The interplay of plant growth regulators in regeneration underscores the need for a comprehensive understanding of the underlying mechanisms. This knowledge is essential to optimize propagation techniques for agricultural and conservation applications. Continued research in this field holds the potential to advance plant biotechnology, enabling efficient regeneration and conservation strategies for diverse plant species (Horstman et al., 2017; Yeboah et al., 2020; Suge, 1984).

3.18 ADVANCES IN TISSUE CULTURE TECHNIQUES

In recent decades, plant biotechnology has achieved significant breakthroughs, particularly in tissue culture methodologies. Central to these advancements is the strategic application of plant hormones, which have proven instrumental in enhancing the efficiency and adaptability of tissue culture techniques.

One of the transformative strides in this field has been the refinement of plant transformation vectors and methodologies. These improvements have greatly advanced the genetic modification of crops, enabling the precise introduction of desirable traits into diverse plant species, including both monocots and dicots. This has been made possible through the development of advanced gene transfer technologies (Rao et al., 2016; Golds et al., 2003).

Somatic embryogenesis has emerged as a cornerstone of tissue culture innovation, particularly in the rapid propagation of horticultural crops. Unlike traditional organogenesis, which often grapples with challenges such as rejuvenation, senescence, and rhizogenesis, somatic embryogenesis enables the efficient production of complete plants with both an apical meristem and primary root. This method has proven invaluable in overcoming bottlenecks in large-scale propagation (Ji et al., 2011).

The advent of in vitro embryogenesis has further revolutionized plant breeding programs by enabling the generation of fully homozygous plants in a single generation (Pandey and Chaudhary 2015; Pandey et.al. 2012). This has accelerated the creation of backcross conversions and genetic marker maps, both of which are essential for developing superior crop varieties (Horstman et al., 2017).

The judicious use of plant hormones, particularly auxins and cytokinins, has played a pivotal role in driving these advancements. By fine-tuning the balance and concentration of these hormones, researchers have optimized key processes such as the initiation and maintenance of somatic embryos and their subsequent conversion into viable plantlets (Saifi et al., 2020; Jaganathan et al., 2018; Horstman et al., 2017; Golds et al., 2003; Pandey and Chaudhary 2014). Despite these achievements, the field continues to face challenges, particularly with recalcitrant species that resist in vitro manipulation. Addressing these challenges will require further optimization of culture conditions, including hormone regimes, to ensure consistent and reliable regeneration outcomes.

3.19 CONCLUSION

The study of PGRs underscores their fundamental role in advancing plant tissue culture methodologies. These natural and synthetic compounds function as essential mediators of cellular processes, orchestrating critical developmental pathways such as organogenesis and somatic embryogenesis. By carefully modulating the concentrations and interactions of PGRs, researchers have achieved precise control over plant growth, facilitating the targeted development of roots, shoots, and embryos.

The inclusion of key PGRs, including auxins, cytokinins, gibberellins, abscisic acid, and ethylene, in culture media has enabled the scientific community to address longstanding challenges in plant regeneration. These growth regulators not only support essential processes like callus induction and plantlet formation but also provide tailored solutions for overcoming species-specific barriers in tissue culture. Their targeted application has catalyzed significant advancements in areas such as crop improvement, large-scale propagation, and the preservation of valuable plant species.

Nevertheless, some challenges remain, including recalcitrance in certain plant species and the necessity of achieving precise hormonal equilibrium for optimal regeneration. Addressing these issues requires further investigation into the molecular mechanisms governing PGR-mediated developmental processes, alongside the development of innovative approaches to enhance the efficiency and reproducibility of tissue culture systems. Such research endeavors hold the potential to broaden the applications of plant biotechnology while contributing to global efforts in sustainable agriculture, food security, and biodiversity preservation.

Therefore, plant growth regulators form the cornerstone of modern plant tissue culture, translating the principle of totipotency into practical applications. Their strategic use continues to drive innovation and sustainability in plant science, solidifying their status as indispensable tools in the quest to unlock new horizons in regeneration and beyond.

REFERENCES

Ali, M. S., & Baek, K. H. (2020). Jasmonic acid signaling pathway in response to abiotic stresses in plants. *International Journal of Molecular Sciences*, *21*(2), 621.

Asami, T., Nakano, T., & Fujioka, S. (2005). Plant brassinosteroid hormones. *Vitamins & Hormones*, *72*, 479–504.

Ascough, G. D., & Fennell, C. (2004). The regulation of plant growth and development in liquid culture. *South African Journal of Botany*, *70*(2), p. 181. Elsevier BV. https://doi.org/10.1016/s0254-6299(15)30234-9

Asif, R., Yasmin, R., Mustafa, M., Ambreen, A., Mazhar, M., Rehman, A., … & Ahmad, M. (2022). Phytohormones as plant growth regulators and safe protectors against biotic and abiotic stress. *Plant Hormones-Recent Advances, New Perspectives and Applications* (pp. 115–130).

Brian, P.W. (1959). Effects of gibberellins on plant growth and development. *Biological Reviews*, *34*, 37–77. https://doi.org/10.1111/j.1469-185X.1959.tb01301.x

Coggins, Jr, C. W., & Lovatt, C. J. (2014). 14 Plant growth regulators. *Citrus Production Manual*, *3539*, 215.

Enders, T. A., & Strader, L. C. (2015). Auxin activity: Past, present, and future. *American Journal of Botany*, *102*(2), 180–196.

Fitch, M. M. M., & Manshardt, R. M. (1990). Somatic embryogenesis and plant regeneration from immature zygotic embryos of papaya (*Carioca papaya* L.). *Plant Cell Reports*, *9*, 320–324.

Fridman, Y., & Savaldi-Goldstein, S. (2013). Brassinosteroids in growth control: How, when and where. *Plant Science*, *209*, 24–31.

Gaj, M. D. (2011). Somatic embryogenesis and plant regeneration in the culture of *Arabidopsis thaliana* (L.) Heynh. Immature zygotic embryos. In Thorpe, T. A., & Yeung, E. C., (Eds.), *Plant Embryo Culture* (pp. 257–265). Berlin, Heidelberg: Springer.

Gaj, M. D. (2004). The role of cytokinins in plant regeneration from tissue cultures. *Plant Cell Reports*, 22(11), 712–720. doi:10.1007/s00299-004-0900-3

Golds, T. J., Davey, M. R., Rech, E., & Power, J. (2003). *Methods of Gene Transfer and Analysis in Higher Plants*. In Humana Press eBooks (p. 341). Humana Press. https://doi.org/10.1385/0-89603-161-6:341

Gordon, S. P., Heisler, M. G., Reddy, G. V., Ohno, C., Das, P., & Meyerowitz, E. M. (2007). Pattern formation During De novo assembly of the Arabidopsis shoot meristem. *Development*, *134*, 3539–3548.

Haberlandt, G. (1902). Cultur Versuche Mit Isolierten Pflanzenzellen Sitz. *Ber Mat Nat ki Kais Akad Wiss, Wien*, 111, 69–92.

Hedden, P., & Sponsel, V. (2015). A century of gibberellin research. *Journal of Plant Growth Regulation*, *34*, 740–760.

Hill, K., & Schaller, G. E. (2013). Enhancing plant regeneration in tissue culture: A molecular approach through manipulation of Cytokinin sensitivity. *Plant Signaling & Behavior*, *8*, e25709.

Horstman, A., Bemer, M., & Boutilier, K. (2017). A transcriptional view on somatic embryogenesis. *Regeneration*, *4*, 201–216.

Huang, W., Xian, Z., Kang, X., Tang, N., & Li, Z. (2015). Genome-wide identification, phylogeny and expression analysis of GRAS gene family in tomato. *BMC Plant Biology*, *15*(1). BioMed Central. https://doi.org/10.1186/s12870-015-0590-6

Iqbal, N., Khan, N. A., Ferrante, A., Trivellini, A., Francini, A., & Khan, M. I. R. (2017). Ethylene role in plant growth, development and senescence: interaction with other phytohormones. *Frontiers in Plant Science*, *8*, 475.

Jaganathan, D., Ramasamy, K., Sellamuthu, G., Shilpha, J., & Venkataraman, G. (2018). CRISPR for crop improvement: An update review [Review of CRISPR for Crop Improvement: An Update Review]. *Frontiers in Plant Science*, *9*. Frontiers Media. https://doi.org/10.3389/fpls.2018.00985

Ji, A., Geng, X., Zhang, Y., & Guo-liang, W. (2011). Advances in somatic embryogenesis research of horticultural plants. *American Journal of Plant Sciences*, *2*(6), 727. Scientific Research Publishing. https://doi.org/10.4236/ajps.2011.26087

Kumar, S., Shah, S. H., Vimala, Y., Jatav H. S., Ahmad, P., Chen, Y. & Siddique, K. H. M. (2022). Abscisic acid: Metabolism, transport, crosstalk with other plant growth regulators, and its role in heavy metal stress mitigation. *Frontiers in Plant Science*, 13, 972856. doi: 10.3389/fpls.2022.972856

Kieber, J. J., & Schaller, G. E. (2014). Cytokinins. *The Arabidopsis Book*, *12*, e0168. doi:10.1199/tab.0168

Li, Y., Han, S., & Qi, Y. (2023). Advances in structure and function of auxin response factor in plants. *Journal of Integrative Plant Biology*, *65*(3), 617–632.

Márquez-López, R. E., Quintana-Escobar, A. O. and Loyola-Vargas, V. M. (2019). Cytokinins, the Cinderella of plant growth regulators. *Phytochemistry Reviews*, *18*(6), 1387–1408.

Miransari, M., & Smith, D. L. (2014). Plant hormones and seed germination. *Environmental and Experimental Botany*, *99*, 110–121.

Muhamad, S. N. S., Ling, A. P. K., & Wong, C. L. (2018). Effect of plant growth regulators on direct regeneration and callus induction from *Sargassum polycystum* C. Agardh. *Journal of Applied Phycology*, *30*, 3299–3310. https://doi.org/10.1007/s10811-018-1649-1

Naqvi, S. S. M. (2001). Plant growth hormones: Growth promoters and inhibitors. In Mohammad Pessarakli (Ed) *Handbook of Plant and Crop Physiology* (pp. 523–548). Boca Raton: CRC Press. eBook ISBN: 9780429208096

Olszewski, N. E., Sun, T. P., & Gubler, F. (2002). Gibberellin signaling: Biosynthesis, catabolism, and response pathways. *The Plant Cell*, *14*(suppl), S61–S80. doi:10.1105/tpc.010092

Oppenheim, J. J. (2001). Cytokines: Past, present, and future. *International Journal of Hematology*, *74*, 3–8.

Orlikowska, T., Nowak, K., & Reed, B. (2017). Bacteria in the plant tissue culture environment. *Plant Cell, Tissue and Organ Culture (PCTOC)*, *128*, 487–508.

Overvoorde, P., Fukaki, H., & Beeckman, T. (2010). Auxin control of root development. *Cold Spring Harbor Perspectives in Biology*, *2*(6), a001537.

Pandey, D. & Chaudhary B. (2014). Oxidative stress responsive *SERK1* gene directs the progression of somatic embryogenesis in cotton (*Gossypium hirsutum* L. cv. Coker 310). *American Journal of Plant Sciences*, *5*(1), 80–102. doi: 10.4236/ajps.2014.51012.

Pandey, D. K., & Chaudhary, B. (2015). Genes and trans-factors underlying embryogenic transition in plant soma-cells. In: Sablok, G., Kumar, S., Ueno, S., Kuo, J., Varotto, C. (eds) *Advances in the Understanding of Biological Sciences Using Next Generation Sequencing (NGS) Approaches*. Springer, Cham. https://doi.org/10.1007/978-3-319-17157-9_10

Pandey, D. K., Singh, A. K., Chaudhary, B. (2012). Boron-mediated plant somatic embryogenesis: A provocative model. *Journal of Botany*, 375829, 9 pages. https://doi.org/10.1155/2012/375829Pierik, R., Sasidharan, R., & Voesenek, L. A. (2007). Growth control by ethylene: Adjusting phenotypes to the environment. *Journal of Plant Growth Regulation*, *26*, 188–200.

Rao, A. Q., Ali, M. A., Khan, M. A. U., Bajwa, K. S., Iqbal, A., Iqbal, T., Shahid, A. A., Nasir, I. A., & Husnain, T. (2016). *Science Behind Cotton Transformation*. In InTech eBooks. https://doi.org/10.5772/64888

Ravnikar, M., Žel, J., Plaper, I., & Špacapan, A. (1993). Jasmonic acid stimulates shoot and bulb formation of garlic in vitro. *Journal of Plant Growth Regulation*, *12*, 73–77.

Ruan, J., Zhou, Y., Zhou, M., Yan, J., Khurshid, M., Weng, W., Cheng, J., & Zhang, K. (2019). Jasmonic acid signaling pathway in plants. *International Journal of Molecular Sciences*, *20*(10), 2479.

Saifi, S. K., Passricha, N., Tuteja, R., Kharb, P., & Tuteja, N. (2020). *In Planta Transformation: A Smart Way of Crop Improvement*. In Elsevier eBooks (p. 351). Elsevier BV. https://doi.org/10.1016/b978-0-12-818581-0.00021-8

Sah, S. K., Reddy, K. R. and Li, J. (2016). Abscisic acid and abiotic stress tolerance in crop plants. *Frontiers in Plant Science*, *7*, 571. doi: 10.3389/fpls.2016.00571

Skoog, F., & Miller, C. (1957). Chemical regulation of growth and organ formation in plant tissues cultured in vitro. In *Symposia of the Society for Experimental Biology* (Vol. 11, pp. 118–130).

Soliman, A. S. (2019). Plant growth hormones. *Cell Growth*, 1, 75–80.

Sosnowski, J., Truba, M., & Vasileva, V. (2023). The impact of auxin and cytokinin on the growth and development of selected crops. *Agriculture*, *13*(3), 724.

Sugimoto, K., Gordon, S. P., & Meyerowitz, E. M. (2011). Regeneration in plants and animals: dedifferentiation, transdifferentiation, or just differentiation? *Trends in Cell Biology*, *21*(4), 212–218. doi:10.1016/j.tcb.2010.12.004

Suman, S. (2017). Plant tissue culture: A promising tool of quality material production with special reference to micropropagation of banana. *Biochemical & Cellular Archives*, *17*(1), 1–26.

Taiz, L., Zeiger, E., Møller, I. M., & Murphy, A. (2015). *Plant Physiology and Development* (No. Ed. 6). Sinauer Associates Incorporated.

Trigiano, R. N., & Gray, D. J. (2016). *Plant Tissue Culture, Development and Biotechnology* (p. 608). Boca Raton, FL: CRC Press.

Ulmasov, T., Murfett, J., Hagen, G., & Guilfoyle, T. J. (1997). Aux/IAA proteins repress expression of reporter genes containing natural and highly active synthetic auxin response elements. *The Plant Cell*, *9*(11), 1963–1971.

Wang, Y., Wu, Q., & Zhang, Y. (2011). Gibberellins in plant regeneration and their application in plant tissue culture. *Plant Growth Regulation*, *63*(1), 1–10. doi:10.1007/s10725-010-9535-4

Weyers, D. B., & Paterson, N. W. (2001a). Plant hormones and environmental responses. *New Phytologist*, *151*(3), 493–511.

Weyers, J. D., & Paterson, N. W. (2001b). Plant hormones and the control of physiological processes. *New Phytologist*, *152*(3), 375–407.

Woodward, A. W., & Bartel, B. (2005). Auxin: Regulation, action, and interaction. *Annals of BOTANY*, *95*(5), 707–735.

Yang, X., & Zhang, X. (2010). Regulation of somatic embryogenesis in higher plants. *Critical Reviews in Plant Science*, *29*, 36–57.

Yeboah, J., Dadzie, A. M., Segbefia, M. A. D., Lowor, S. T., Agene, V. N., Osei-Akoto, S., Owusu-Ansah, F., Banful, B., & Atuah, L. (2020). Vegetative propagation of cashew (Anacardium occidentale L.) by softwood grafting in Ghana. *Journal of Agricultural Science*, *12*(8), p. 257. Canadian Center of Science and Education. https://doi.org/10.5539/jas.v12n8p257

Zhang, L., DeGennaro, D., Lin, G., Chai, J., & Shpak, E. D. (2021). Erecta family signaling constrains Clavata3 and Wuschel to the center of the shoot apical meristem. *Development*, *148*, dev189753.

4 Somatic Embryogenesis
A Key Pathway of Regeneration

Mahwish Firdous, Aakriti Gupta, Poonam Kumari Yadav, and Dhananjay Kumar Pandey

4.1 INTRODUCTION

Somatic embryogenesis (SE) is the process where specialized, nonreproductive somatic cells undergo remarkable developmental reprogramming to form structures that are analogous to the embryos produced through sexual reproduction (Zimmerman, 1993). These embryo-like structures can then develop into fully functional plants. This phenomenon underscores the inherent totipotency of plant cells, a concept first proposed by Gottlieb Haberlandt in the early 20th century and later demonstrated through various experimental studies (Fehér, 2019). The capacity of somatic cells to revert to an embryonic state and regenerate complete plants has significant implications for plant science, agriculture, and biotechnology, enabling the mass production of economically important cultivars (Bidabadi & Jain, 2020). The foundational work in this field has paved the way for extensive research into the mechanisms, applications, and optimization (Pandey & Chaudhary, 2015; Goldberg et al., 1994; Babak et al., 2008; Tisserat et al., 1979; Fehér, 2015; Rocha & Dornelas, 2013; Méndez-Hernández et al., 2019).

4.2 DIFFERENCE BETWEEN SOMATIC AND ZYGOTIC EMBRYOGENESIS

SE is a type of induced plant cell totipotency where embryos develop from somatic or vegetative cells without fertilization. Molecular genetic studies have demonstrated that ectopic expression of specific transcription factors expressed by embryos and meristems, or the loss of certain chromatin-modifying proteins, can induce spontaneous SE (Karami et al., 2009; Rocha & Dornelas, 2013; Salaün et al., 2021; Horstman et al., 2017). *In vitro* SE can be induced by subjecting explants to stress or growth regulator treatments (Horstman et al., 2017). Somatic embryos are clones because they are created without fertilization and share the same genetic makeup as their parent tissue (Magnani et al., 2017). This technique relies on the concept of cellular totipotency, which is the capacity of a single cell to divide and develop into a complete and functional plant (Salaün et al., 2021). Under appropriate induction

DOI: 10.1201/9781032702391-4

conditions, a subset of cells can generate embryogenic cells that lead to the initiation of the somatic embryo (Vogel, 2005). SE can originate from a single cell or a collection of cells and can have a multicellular origin (Harada et al., 2010). In contrast to zygotic embryogenesis (ZE), which begins with a single cell and results in the production of a globular embryo with a fixed number of cells, SE begins with a single cell or a group of cells and creates a globular structure with a variable number of cells (Harada et al., 2010). ZE in higher plants refers to the developmental stage during which the zygote undergoes a sequence of differentiation events leading to the formation of a mature embryo, guided by the hormone auxin. Partitioning events along the apical-basal axis establish the major embryonic organs and shoot and root apical meristems (Smet et al., 2010).

4.3 PATHWAY OF SE

The processes following the induction of SE in a cultured explant are well described at the histological level and are known as indirect and direct embryogenesis. However, it can be challenging to distinguish between the two, and they often occur simultaneously on the same explant (Horstman et al., 2017).

4.3.1 Indirect SE

Indirect SE involves the development of a callus as a stage preceding the formation of somatic embryos. It is important to note that not all plant cells possess this morphogenic ability. Therefore, identifying the kinds of stimuli that trigger this response has proven difficult (Karami et al., 2009). Explants dedifferentiate to generate callus, from which cells differentiate to form somatic embryos. Due to its early amorphous form, callus was and is commonly referred to as "undifferentiated" or "dedifferentiated."

4.3.2 Direct SE

Direct SE is the ability of explants to stimulate somatic embryos directly without a callus phase. In this system, single or multiple cells in single or multiple cell layers divide and bulge to form embryos that are morphologically identifiable without additional treatment. This method exhibits more regular and less prolific compact cell divisions (Horstman et al., 2017; Harada et al., 2010). Although somatic embryos formed from direct and indirect paths share a similar morphology, the longer tissue culture period of indirect SE embryos generally results in genome-level alterations or somaclonal diversity (Miguel & Marum, 2011). Furthermore, certain plants have the ability to produce somatic embryos both directly and indirectly. Examples include cucumbers, orchids, walnuts, and *Chrysanthemum morifolium* (Ji et al., 2011).

4.4 *IN VIVO* AND *IN VITRO* SE

In vivo SE refers to the process where somatic cells within plant tissue are stimulated to undergo embryogenesis, resulting in the development of somatic embryos within the entire plant tissue. Unlike *in vitro* SE, which takes place in culture conditions, *in vivo* SE usually involves using different physiological and biochemical stimuli to induce embryogenic potential in somatic cells within the plant tissues. These cues may consist of genetic modifications, stress situations, and adjustments to growth regulators (Rocha & Dornelas, 2013; Altamura et al., 2016). Examples of *in vivo* SE include nucellar polyembryony in citrus, somatic embryos produced on the cotyledons of certain seedlings, and the formation of somatic embryos on the surface of leaves (Tapingkae et al., 2012; Harada et al., 1998). The somatic cells undergo morphological and biochemical changes similar to those that occur during ZE. This includes the development of somatic embryos, the creation of embryogenic callus, and ultimately the transformation of somatic embryos into fully grown plants. Plant biotechnology may find use for *in vivo* SE in mass propagation of elite plant genotypes, genetic modification, and the creation of artificial seeds for effective distribution and conservation of germplasm (McKersie & Brown, 1996; Zimmerman, 1993). *In vitro* SE refers to the method of stimulating the creation of embryos from somatic (nonreproductive) cells in a controlled laboratory setting. Rather than using fertilized egg cells, this method stimulates the growth of embryos from cells that are not typically engaged in embryogenesis, such as leaf or root cells (Dodeman et al., 1997; Rocha & Dornelas, 2013). *In vitro* SE is a potent method in plant biotechnology for large-scale plant development, particularly in horticulture, forestry, and crops. It allows for the production of genetically identical plants, or clones, from a single parent plant and rapid propagation of plants exhibiting desired traits. Furthermore, it finds application in genetic engineering, plant breeding, and the preservation of rare or endangered plant species (Xiong et al., 2022).

4.5 FACTORS AFFECTING SE

4.5.1 Physiological and Physical Factors

SE in plants is influenced by several physiological factors, including genotype, age, and developmental stage of the explant, endogenous hormones, nutrient availability, physical environment, oxygen availability, and stress conditions (Altamura et al., 2016; Fehér, 2015; Méndez-Hernández et al., 2019). The genetic composition of plant tissue plays a significant role in SE, as some genotypes respond better to somatic embryo induction and development than others. The tissue initiated for SE affects its efficiency, with younger tissues typically showing greater potential for developing into embryos (Zimmerman, 1993). The concentration and balance of endogenous hormones like auxins, cytokinins, and abscisic acid (ABA) are crucial for SE. Auxins and cytokinins are frequently used in culture media to enhance embryogenic responses, while ABA may impair development (Carneros et al., 2023). The composition and concentration of nutrients in the culture medium influence the embryogenic potential of plant tissues. Somatic embryos require specific nutrient levels, such as carbon, nitrogen, minerals, and vitamins, for their proper development and maturation (Ochatt & Revilla, 2016). Several physical factors significantly influence SE in plants. Light, temperature, and oxygen levels are critical factors that can affect the initiation, development, and maturation of somatic embryos (Nolan & Rose, 2010; Zimmerman, 1993). Specific light regimes and optimal temperature ranges are necessary for the formation of embryogenic cultures and the growth of somatic embryos (Merkle et al., 2006). The physical environment of the culture system, including the types of culture vessels and culture media consistency, can also influence the development of somatic embryos. Additionally, oxidative stress or mechanical stress can stimulate embryogenic responses in plant tissues, potentially leading to the formation of somatic embryos (Nolan & Rose, 2010; Karami et al., 2009; Ochatt & Revilla, 2016). Exposing plant tissues to specific electromagnetic fields can also stimulate embryogenic responses and improve the efficiency of somatic embryo induction. Moreover, microgravity conditions, such as those experienced in space, can influence the growth patterns and morphogenesis of somatic embryos, resulting in unique developmental outcomes (Theimer et al., 1986; Khai et al., 2021).

4.5.2 Hormonal Factors

SE in plants is significantly influenced by various hormones, including auxins, cytokinins, ABA, gibberellins, and ethylene. Auxins, such as indole-3-acetic acid and 2,4-dichlorophenoxyacetic acid, are crucial in SE. They promote cell division, elongation, and differentiation, facilitating the formation of somatic embryos. High auxin concentrations are often used to induce embryogen (Karami et al., 2022). Similarly, cytokinins play a role in cell division and differentiation and work synergistically with auxins to promote SE, enhancing the formation of embryogenic callus and the development of somatic embryos (Asghar et al., 2022; Avilez-Montalvo et al., 2022) ABA regulates somatic embryo maturation and stress responses. It is involved in processes such as embryo dormancy, desiccation tolerance, and the development of storage reserves (Méndez-Hernández et al., 2019). The presence of ABA can influence the formation and quality of somatic embryos (Tian & Brown, 2000). Gibberellins, such as gibberellic acid, are involved in cell elongation and growth (Vishal & Kumar, 2018). They can affect SE by influencing cell differentiation and promoting the development of embryogenic tissues (Li, 2023). Moreover, ethylene is a gaseous hormone that can impact SE. It affects various developmental processes, including cell division and differentiation, and can modulate the induction and development of somatic embryos (Mira et al., 2015). The intricate interplay between these hormones and their precise regulation are crucial for the successful induction, development, and maturation of somatic embryos in plants (Méndez-Hernández et al., 2019; Horstman et al., 2017; Karami et al., 2009).

4.5.3 Stress Factors

Biotic and abiotic stress factors play a crucial role in regulating the process of SE in plants (Fehér, 2015). These environmental cues can profoundly influence the cellular responses and the embryogenic potential of the plant tissues undergoing this complex developmental program (Gulzar et al., 2020). Regarding biotic stress factors, pathogenic microorganisms, herbivorous pests, and competing organisms can significantly disrupt the normal physiological functions and signaling pathways required for the induction and progression of SE (Ochatt & Revilla, 2016). The presence of these biotic stressors can trigger various defense responses in the plant cells, which may subsequently alter their embryogenic competence and compromise their ability to undergo the complex developmental transitions leading to the formation of somatic embryos (Fehér, 2015). Similarly, abiotic stress factors, such as variations in temperature, light intensity, humidity, salinity, pH, and nutrient availability, can also have a substantial impact on SE (Fehér, 2015). These environmental cues can either stimulate or inhibit the embryogenic responses in the plant tissues, depending on the intensity and duration of the stress exposure. Furthermore, the abiotic stress conditions can directly influence the growth, differentiation, and maturation of the developing somatic embryos, ultimately affecting their quality, viability, and overall developmental success (Fehér, 2015; Pandey et al., 2012; Ochatt & Revilla, 2016).

4.6 CELLULAR AND MOLECULAR VIEW OF SE

4.6.1 Phytohormone Mediated Control

SE is a complex process involving various proteins and genes regulating cell cycle, hormone signaling, chromatin remodeling, and metabolic pathways (Fehér, 2015). These proteins and genes can be classified based on their roles in different stages of embryo induction, development, and maturation (Table 4.1). Hormonal, genetic, and epigenetic controls regulate SE (Horstman et al., 2017; Karami et al., 2009). Hormonal control, particularly through the regulation of auxin, is a fundamental aspect of SE in plants (Wójcik et al., 2020). Auxin is a key phytohormone that governs various stages of plant development, including the acquisition of embryogenic competence, cell division, and differentiation, all of which are essential for

TABLE 4.1
Developmental Stages and Hormonal Requirements in Somatic Embryogenesis

Stage	Key Developmental Features	Dominant Hormones	Role of Hormones
Induction	Dedifferentiation of somatic cells; callus formation	Auxins (e.g., 2,4-D)	Promotes cell division and dedifferentiation
Globular	Formation of globular embryo structures	Auxins, cytokinins	Balances cell proliferation and embryonic patterning
Heart-shaped	Establishment of embryo polarity	Gibberellins, cytokinins	Supports elongation and axis formation
Torpedo	Elongation and differentiation of embryo	Abscisic acid (ABA)	Initiates maturation and stress tolerance
Maturation	Embryo develops cotyledons and stores nutrients	ABA, reduced auxins	Ensures desiccation tolerance and dormancy
Germination	Conversion of embryo into a viable plantlet	Reduced ABA, gibberellins	Facilitates resumption of growth and organogenesis

the successful initiation and progression of somatic embryos (Tang et al., 2020). The orchestration of auxin signaling is mediated by several critical proteins, including PIN-FORMED (PIN) proteins and AUXIN RESPONSE FACTORS (ARFs), which together facilitate the precise spatial and temporal regulation of gene expression necessary for SE (Fehér, 2015). PIN proteins play a crucial role in the polar transport of auxin, effectively establishing and maintaining auxin gradients within plant tissues (Zhou & Luo, 2018). These gradients are vital for the proper patterning and development of somatic embryos, as they dictate the localized concentrations of auxin that influence cell fate decisions (Su & Zhang, 2009). The directional flow of auxin, governed by PIN proteins, ensures that specific regions of the tissue acquire the necessary auxin concentrations to trigger embryogenic processes (Möller & Weijers, 2009). This spatial regulation is essential for the formation of distinct embryonic structures and the overall organization of the developing somatic embryo. In conjunction with PIN proteins, ARFs are key regulators that mediate the transcriptional responses to auxin (Wójcikowska & Gaj, 2017). ARFs bind to auxin-responsive elements in the promoters of target genes, modulating their expression in response to fluctuating auxin levels (Li et al., 2016). Through this mechanism, ARFs control the expression of genes involved in critical processes such as cell division, differentiation, and the establishment of embryogenic competence (Li et al., 2016). The activity of ARFs ensures that cells respond appropriately to auxin signals, enabling the coordinated development of the somatic embryo from a population of undifferentiated somatic cells (Guilfoyle & Hagen, 2007). High concentrations of auxin are frequently employed *in vitro* to induce SE. This exogenous application of auxin mimics the endogenous hormonal environment required for the reprogramming of somatic cells into an embryogenic state. During this process, auxin-responsive genes, including key regulators such as *LEAFY COTYLEDON 1* (*LEC1*), *LEAFY COTYLEDON 2* (*LEC2*), and *FUSCA3* (*FUS3*), are activated (Tang et al., 2020; Wójcikowska et al., 2013; Stone et al., 2023a; Wójcik et al., 2020). These transcription factors play crucial roles in the development and maturation of somatic embryos, functioning downstream of auxin signaling to drive the expression of genes necessary for embryogenesis.

4.6.2 Genetic Control

Several transcription factors and genes play vital roles in regulating gene expression during SE and embryonic development.

4.6.2.1 *LEC1, LEC2,* and *FUS3*

LEC1, *LEC2*, and *FUS3* are pivotal transcription factors that govern the intricate processes of SE in plants (Liu et al., 2021; Braybrook et al., 2006; Salaün et al., 2021; Ledwoń & Gaj, 2011). These transcriptional regulators are central to the acquisition of embryogenic competence in somatic cells and the subsequent development and maturation of somatic embryos. They operate through a coordinated regulatory network, modulating gene expression to drive key cellular processes such as cell division, differentiation, and maturation. *LEC1*, a member of the CCAAT-binding transcription factor complex, is crucial for the initial induction of the embryogenic program (Lee et al., 2003; Braybrook et al., 2006). *LEC1* activates a broad spectrum of genes that are essential for establishing embryonic identity in somatic cells, thereby facilitating their transition into an embryogenic state (Junker & Bäumlein, 2012; Braybrook et al., 2006; Ledwoń & Gaj, 2011). This transition involves extensive reprogramming, including the activation of genes that encode storage proteins, lipid bodies, and other embryogenesis-associated components. *LEC1*'s role extends beyond merely initiating embryogenesis; it also maintains the embryonic identity of cells by suppressing differentiation into non-embryonic cell types (Junker & Bäumlein, 2012; Salaün et al., 2021). This maintenance is critical in ensuring that the somatic cells retain their capacity to develop into fully functional embryos. *LEC2*, a B3 domain transcription factor, functions downstream of *LEC1* and plays a critical role in reinforcing the embryogenic program (Stone et al., 2023b). *LEC2* is particularly important for the regulation of auxin biosynthesis and signaling pathways, which are indispensable for the initiation and maintenance of SE (Braybrook et al., 2006). Auxin is a key phytohormone that modulates cell division, expansion, and differentiation, and *LEC2* ensures the proper hormonal environment by controlling the expression of genes involved in auxin homeostasis (Luo et al., 2018). Additionally, *LEC2* influences lipid metabolism and storage, which are vital for providing the energy and structural components required for embryo development. The precise regulation of *LEC2* expression is crucial for the successful progression of embryogenesis, as it fine-tunes the cellular environment to support the developing embryos (Braybrook et al., 2006). *FUS3*, another B3 domain transcription factor, plays a dominant role during the later stages of somatic embryo development, particularly in the maturation phase (Tsuchiya et al., 2023). *FUS3* regulates genes involved in desiccation tolerance, storage reserve accumulation, and the establishment of dormancy, which are essential for the proper maturation of somatic embryos (Luerssen et al., 1998; Nambara et al., 2000). By promoting the accumulation of storage proteins and lipids, *FUS3* ensures that the embryos are adequately equipped to survive the transition from an embryonic to a post-embryonic state (Tsuchiya et al., 2023). *FUS3* also interacts with ABA signaling pathways, integrating environmental cues with the internal developmental program to coordinate stress responses and maturation processes (Wang & Perry, 2013; Gulzar et al., 2020). The regulation of these processes by *FUS3* is critical for the development of viable, desiccation-tolerant embryos that can withstand adverse environmental conditions. The coordinated action of *LEC1*, *LEC2*, and *FUS3* forms a robust regulatory network that ensures the successful execution of SE (Gulzar et al., 2020). *LEC1* initiates the embryogenic program and maintains embryonic identity, *LEC2* reinforces this program by modulating auxin signaling and lipid metabolism, and *FUS3* oversees the maturation and final preparation of the embryo for post-embryonic life (Gulzar et al., 2020). The intricate interplay among these transcription factors underscores their central role in the regulation of SE, making them key targets for biotechnological applications aimed at enhancing the efficiency of SE. Manipulating the expression of *LEC1*, *LEC2*, and *FUS3* can significantly improve the induction and development of somatic embryos, offering valuable tools for plant propagation, genetic transformation, and conservation efforts (Guo et al., 2013; Liu et al., 2018).

4.6.2.2 *BABY BOOM (BBM)*

BBM plays a central role in the initiation of somatic embryos from somatic cells, driving both the reprogramming of these cells to an embryogenic state and their subsequent proliferation and differentiation into fully developed embryos (Boutilier et al., 2002). The primary function of *BBM* in SE is the induction of embryogenic competence in somatic cells. *BBM* initiates this process by activating a specific set of target genes that reprogram somatic cells, enabling them to transition from a differentiated state to an embryogenic state (Kulińska-Łukaszek et al., 2012). This reprogramming is characterized by the activation of genes that promote cell division, leading to a marked increase in cell proliferation. This proliferative activity is a crucial early step in SE, as it generates a population of undifferentiated cells that serve as the foundation for the formation of somatic embryos. In addition to its role in cell proliferation, *BBM* is also essential for the development and differentiation of somatic embryos (Kulińska-Łukaszek et al., 2012). *BBM* regulates the expression of genes involved in cell fate determination and organogenesis, ensuring that the proliferating embryogenic cells undergo the correct differentiation pathways to form the various tissues and organs of the developing embryo (Kulińska-Łukaszek et al., 2012). This regulation is critical for the proper morphological and physiological development of somatic embryos, enabling them to transition through the various stages of embryogenesis to become viable plantlets (Table 4.1). The function of *BBM* is further modulated through its interaction with hormonal signaling pathways, particularly those involving auxins and cytokinins (Méndez-Hernández et al., 2019; Boutilier et al., 2002). These hormones are pivotal in maintaining the balance between cell proliferation and differentiation during SE. *BBM* influences the expression of genes involved in the biosynthesis, transport, and signaling of these hormones, thereby integrating hormonal cues with the transcriptional regulation of embryogenesis. This integration ensures that the embryogenic cells remain responsive to endogenous and exogenous signals, which is necessary for the sustained progression of somatic embryo development (Horstman et al., 2017). The strategic role of *BBM* in regulating SE makes it a key target for enhancing SE in plant biotechnology (Horstman et al., 2017). Overexpression of *BBM* has been demonstrated to significantly improve the efficiency of SE, particularly in species that are typically recalcitrant to *in vitro* culture or SE induction (Boutilier et al., 2002; Chen et al., 2022). By promoting the proliferation of embryogenic cells and guiding their differentiation, *BBM* can be leveraged to enhance the production of somatic embryos, thereby improving outcomes in clonal propagation, genetic transformation, and the conservation of plant genetic resources. Therefore, we can consider *BBM* as a master regulator of SE, with a multifaceted role that encompasses the initiation of embryogenic competence, promotion of cell proliferation, and regulation of embryo development (Boutilier et al., 2002). Its interaction with hormonal pathways further enhances its regulatory capacity, making *BBM* an essential factor in the successful execution of SE and a valuable tool in plant biotechnology applications.

4.6.2.3 *WUSCHEL (WUS)* and *WUSCHEL-RELATED HOMEOBOX (WOX)* Genes

WUSCHEL (WUS) and *WUSCHEL-RELATED HOMEOBOX (WOX)* genes are crucial transcription factors that regulate key aspects of plant development, including the maintenance of stem cell populations and the induction of SE (Zuo et al., 2002).

These genes encode homeodomain proteins that are integral to the reprogramming of somatic cells, facilitating their transition into an embryogenic state and guiding the subsequent development of somatic embryos. *WUS*, a central member of the *WOX* gene family, is primarily known for its role in maintaining stem cell niches within the shoot apical meristem (SAM). In the context of SE, *WUS* functions as a master regulator that initiates and sustains the embryogenic potential of somatic cells (Graaff et al., 2009; Zuo et al., 2022). Upon activation, *WUS* induces the expression of a network of genes involved in cell proliferation and meristematic activity, effectively reprogramming differentiated somatic cells into a pluripotent, embryogenic state. This reprogramming is essential for the initiation of somatic embryos, as it establishes a population of undifferentiated cells capable of giving rise to the various tissues and organs of the embryo (Fambrini et al., 2022; Gallois et al., 2004; Su et al., 2009). *WUS* plays a critical role in maintaining the balance between cell proliferation and differentiation during SE. It prevents the premature differentiation of embryogenic cells, thereby ensuring that these cells retain their capacity for continued proliferation and developmental plasticity (Su et al., 2009). This regulation is particularly important in the early stages of SE, where the establishment of a stable embryogenic cell population is necessary for the subsequent formation of organized embryonic structures (Cao, 2013). The expression of *WUS* is tightly regulated and is often induced by exogenous phytohormones such as auxins and cytokinins, which are commonly used in tissue culture to trigger SE (Fambrini et al., 2022). Once expressed, *WUS* coordinates the activation of downstream genes involved in cell division, pattern formation, and organogenesis, ensuring the orderly progression of somatic embryos through their developmental stages (Table 4.1) (Zuo et al., 2022). The ability of *WUS* to maintain a stem cell-like state in embryogenic cells is crucial for the successful establishment and maintenance of embryonic structures during SE. The *WUSCHEL-RELATED HOMEOBOX (WOX)* gene family, which includes *WUS*, comprises a group of transcription factors that are involved in various stages of embryogenesis (Zuo et al., 2002; Zhou et al., 2018). *WOX* genes play complementary roles to *WUS* in regulating embryonic development, from the initial acquisition of embryogenic competence to the differentiation and maturation of somatic embryos. For instance, *WOX2* and *WOX5* are involved in early embryogenesis, where they contribute to the establishment of the apical-basal axis and the specification of embryonic cell fates (Chen et al., 2009; Zhou et al., 2018). These *WOX* genes operate within a regulatory network that governs the spatial and temporal expression of key developmental genes, ensuring the correct patterning and differentiation of the embryo. *WOX* genes also play a role in maintaining the pluripotency of embryogenic cells, similar to *WUS*'s role in stem cell maintenance within the SAM (Wu et al., 2007). By modulating the expression of genes associated with meristem identity, cell division, and developmental timing, *WOX* genes ensure that embryogenic cells remain capable of generating all the necessary cell types for complete embryo development. This regulatory activity is critical for the transition from an undifferentiated embryogenic state to a fully differentiated embryo, capable of forming a viable plantlet (Gambino et al., 2010). The roles of *WUS* and *WOX* genes in SE highlight their significance in plant developmental biology and their potential applications in plant biotechnology.

4.6.2.4 *WIND* Genes

WOUND-INDUCED DEDIFFERENTIATION (WIND) genes are critical regulators in plant SE, particularly in the context of cellular reprogramming following wounding or stress (Iwase et al., 2018). These genes encode transcription factors that are pivotal in initiating the dedifferentiation of somatic cells, thereby enabling them to regain pluripotency and enter an embryogenic state. The *WIND* gene family plays a significant role in facilitating the initial stages of SE by responding to injury and other external stimuli, which often serve as triggers for the embryogenic process (Iwase et al., 2021). *WIND* genes are rapidly activated in response to physical damage or stress conditions, initiating a cascade of molecular events that lead to cellular differentiation (Iwase et al., 2021). This process is crucial for the reprogramming of differentiated somatic cells back into a pluripotent state, a prerequisite for the initiation of SE. *WIND* genes achieve this by upregulating the expression of other key regulators of embryogenesis, including transcription factors such as *WUSCHEL (WUS)* and BABY BOOM (*BBM*), which are essential for maintaining the embryogenic competence of cells (Ikeda & Ohme-Takagi, 2014). The activation of *WIND* genes thus serves as an early molecular marker of the dedifferentiation process that precedes the formation of somatic embryos in response to wounding by promoting cell proliferation and the formation of callus tissue (Iwase et al., 2016; Ikeda & Ohme-Takagi, 2014; Iwase et al., 2011). The wound-induced callus formation is often the first step in the SE pathway, providing a population of cells that can be reprogrammed into somatic embryos under the right conditions (Pandey & Chaudhary, 2015). *WIND* genes regulate the expression of genes involved in cell cycle progression, enabling the rapid proliferation of these dedifferentiated cells. In addition to their role in promoting cell proliferation, *WIND* genes are also involved in the hormonal signaling pathways that are critical for SE (Iwase et al., 2015; Salaün et al., 2021). Furthermore, *WIND* genes also play a role in enhancing the plasticity of somatic cells, allowing them to respond flexibly to external stimuli and re-enter the cell cycle (Ikeda & Ohme-Takagi, 2014). This plasticity is essential for the successful reprogramming of cells into an embryogenic state, as it allows somatic cells to adopt new identities and developmental trajectories based on the environmental and physiological context. The ability of *WIND* genes to induce such plasticity is a key factor in their role as mediators of SE, particularly in response to stress or injury (Iwase et al., 2011; Iwase et al., 2015; Iwase et al., 2018). The importance of *WIND* genes in SE has significant implications for plant biotechnology. By manipulating the expression of *WIND* genes, it might be possible to enhance the efficiency of SE, particularly in species or tissues that are otherwise recalcitrant to *in vitro* culture and regeneration. Overexpression of *WIND* genes can lead to increased rates of dedifferentiation and callus formation, thereby improving the overall success of SE protocols (Iwase et al., 2011). This makes *WIND* genes valuable targets for biotechnological applications aimed at improving clonal propagation, genetic transformation, and the conservation of plant genetic resources as it ensure the successful reprogramming of somatic cells into embryogenic cells. Their role in enhancing cellular plasticity and responsiveness to external stimuli makes them critical components of the molecular framework underlying SE and valuable tools for advancing plant biotechnology.

4.6.2.5 AGAMOUS-LIKE 15 (AGL15)

AGAMOUS-LIKE 15 (AGL15) is a MADS-box transcription factor that plays a central role in the molecular regulation of SE in plants (Joshi et al., 2022; Zheng et al., 2013). SE is a process in which differentiated somatic cells are reprogrammed to form embryos, independent of fertilization. This pathway is critical for plant biotechnology applications, including clonal propagation, genetic transformation, and conservation of plant genetic resources. AGL15 is highly expressed during the early stages of both somatic and ZE (Thakare et al., 2008; Heck et al., 1995). It functions as a key regulatory factor in the transition of somatic cells from a vegetative to an embryogenic state. AGL15 achieves this by modulating the expression of a suite of target genes associated with embryogenesis. Specifically, it influences the transcriptional landscape necessary for the acquisition of embryogenic competence, thereby initiating the formation of embryonic structures from somatic cells (Zheng et al., 2013; Wang et al., 2004; Joshi et al., 2022; Heck et al., 1995). This reprogramming involves a complex interplay of genetic and epigenetic mechanisms, which are orchestrated by AGL15 to ensure the proper establishment of the embryogenic pathway (Joshi et al., 2022). Studies have demonstrated that the MADS-box transcription factor AGAMOUS-LIKE 15 and the microRNA miR156 play a central role in regulating the induction of SE. Additionally, AGL15 has been reported to modulate the expression of the MIR156 genes within plant systems (Nowak et al., 2020). The function of AGL15 is not limited to direct gene regulation; it also interacts with other signaling pathways, particularly those involving phytohormones like auxins, which are pivotal in the induction and maintenance of SE (Joshi et al., 2022; Zheng et al., 2016). AGL15 has been shown to modulate the expression of genes involved in auxin biosynthesis, signaling, and response, thereby integrating hormonal cues into the regulatory network that governs embryogenesis (Joshi et al., 2022; Zheng et al., 2016). Furthermore, AGL15 interacts with other transcription factors and epigenetic regulators, facilitating a coordinated and robust embryogenic response. These interactions highlight the role of AGL15 as a central node in a complex regulatory network that ensures the proper progression of SE. From a biotechnological perspective, the manipulation of AGL15 expression represents a powerful tool for enhancing SE efficiency, particularly in species that are recalcitrant to *in vitro* culture techniques. Overexpression of AGL15 has been shown to increase the frequency and quality of somatic embryos, thereby improving the efficiency of clonal propagation and genetic transformation protocols. Moreover, AGL15 serves as a molecular marker for the identification and selection of embryogenic cultures, streamlining the process of somatic embryo production in both research and commercial applications. Therefore, AGL15 acts as a master regulator of SE, exerting control over the key developmental transitions and interacting with a network of other regulatory factors to ensure the successful reprogramming of somatic cells into embryogenic cells. Its role in the regulation of embryogenic competence, somatic embryo maturation, and interaction with hormonal pathways underscores its importance in plant developmental biology and its potential applications in plant biotechnology.

4.6.2.6 *SERK*

SERK genes are upregulated in response to specific induction signals, such as the presence of auxins and cytokinins in the culture medium (Méndez-Hernández et al., 2019). This upregulation primes somatic cells to acquire competence for embryogenesis. SERK proteins, located on the cell membrane, function as receptors that perceive and transduce extracellular signals into intracellular responses (Pandey & Chaudhary, 2014a,b). This signaling cascade subsequently activates downstream transcription factors and genes involved in cell differentiation and embryogenesis (Pandey & Chaudhary, 2015). Through their signaling roles, *SERK* genes reprogram somatic cells by altering gene expression patterns, which is crucial for cell dedifferentiation and the initiation of the embryogenic pathway (Méndez-Hernández et al., 2019). Moreover, *SERK* genes play a continuous role in maintaining the embryogenic state of cells by modulating the expression of other regulatory genes, ensuring proper embryogenesis progression (Pandey & Chaudhary, 2014a). Key transcription factors activated downstream of *SERK* signaling including *WUSCHEL (WUS)* and LEAFY COTYLEDON (LEC), which are essential in establishing and sustaining the embryogenic program (Méndez-Hernández et al., 2019).

4.6.3 Epigenetic Control

Epigenetic regulation, encompassing DNA methylation, histone modifications, and chromatin remodeling, is essential for SE, as it governs the expression of genes critical for the development of somatic embryos (Li et al., 2024). DNA methylation is a fundamental epigenetic modification that plays a pivotal role in regulating gene expression (Moore et al., 2012). The patterns of DNA methylation can significantly influence the embryogenic potential of somatic cells, thereby affecting their capacity to develop into somatic embryos (Chakrabarty et al., 2003). Histone modifications, including acetylation, methylation, and phosphorylation, further contribute to the regulation of chromatin structure and gene expression (Frost et al., 2024). These modifications are integral to the induction and development of somatic embryos, as they impact the accessibility of specific genes required for embryogenic competence (Wójcikowska et al., 2020). In addition to these modifications, chromatin remodeling complexes are vital for regulating the accessibility of DNA to transcription factors and other regulatory proteins (Zhang et al., 2012). These complexes facilitate the reprogramming of somatic cells, enabling the induction of SE by altering the chromatin structure to promote or inhibit the expression of key developmental genes. One notable chromatin remodeling factor is PICKLE (PKL), which has been shown to play a critical role in regulating SE by influencing the expression of genes necessary for establishing embryogenic competence and guiding the development of somatic embryos (Ogas et al., 1997; Xu & Huang, 2014). Similarly, VAL1 functions as a transcriptional repressor that maintains the epigenetic state of somatic cells, thereby influencing their transition from a differentiated to an embryonic growth state (Liang et al., 2022; Xu & Huang, 2014). Moreover, microRNAs (miRNAs), which are small noncoding RNAs, are essential players in the posttranscriptional regulation of gene expression during SE (Wójcikowska et al., 2020). These molecules modulate the expression of target genes involved in key processes such as cell division, differentiation, and development, thus fine-tuning the regulatory networks that govern embryonic development (Siddiqui et al., 2019; Ramakrishnan et al., 2023). Collectively, the interplay of these epigenetic mechanisms ensures that somatic cells can effectively reprogram and develop into somatic embryos, highlighting the intricate regulatory landscape that underlies SE in plants.

4.7 APPLICATIONS

SE is a pivotal process in plant biotechnology and agriculture, offering a multitude of applications that range from large-scale propagation to advanced genetic research (Table 4.2). One of its most significant uses is the mass production of selected genotypes, enabling the rapid propagation of elite plant varieties with desirable traits (Bornman, 1993; Meucci et al., 2024). This capability is particularly advantageous for the commercial propagation of high-value crops, where consistency in plant quality and performance is essential (Meucci et al., 2024). Somatic embryos can develop into plantlets with high efficiency, making this technique a cornerstone of modern plant tissue culture practices. In addition to its role in propagation, SE facilitates the production of artificial seeds (Gray et al., 1991). These encapsulated somatic embryos can be directly sown in the field, providing a practical and cost-effective method for large-scale plant propagation. The use of artificial seeds simplifies the handling and distribution of planting material, particularly in regions where traditional propagation methods are limited by environmental or logistical constraints (Rihan et al., 2017). SE also plays a critical role in haploid breeding, where it is used to generate plants with a single set of chromosomes. Haploid plants are invaluable in genetic studies, mutation breeding, and the development of homozygous lines, which are

TABLE 4.2
Applications of Somatic Embryogenesis in Plant Biotechnology

Application	Description	Example Plants	Significance
Clonal propagation	Rapid multiplication of genetically uniform plants	Banana, sugarcane, oil palm	Ensures large-scale production of elite genotypes
Genetic engineering	Introduction of new traits via somatic embryos	Rice, maize, cotton	Facilitates the development of transgenic crops with improved traits
Synthetic seed production	Encapsulation of somatic embryos for artificial sowing	Alfalfa, carrot	Enables easy storage, transport, and large-scale planting
Germplasm conservation	Long-term storage of somatic embryos for conservation	Endangered medicinal plants	Preserves genetic diversity and ensures availability for future breeding programs
Pathogen-free plants	Regeneration of disease-free plants from infected tissues	Potato, grapevine	Helps in managing plant diseases and improving crop health

essential for plant breeding programs aimed at producing uniform and stable crop varieties (Dunwell, 2010). The ability to produce haploids through SE accelerates the breeding process, enabling the rapid development of new cultivars with improved traits (Shen et al., 2023). A key advantage of SE is its capacity for asexual propagation, allowing for the production of genetically uniform clones (Gray et al., 1991). This is particularly useful in horticulture and forestry, where maintaining the genetic integrity and uniformity of plant populations is critical for commercial success. Clonal propagation through SE ensures that the desirable traits of a given genotype are consistently expressed across all propagated plants. Moreover, somatic embryos are highly amenable to genetic modification, owing to their limited somaclonal variation and high receptivity to foreign DNA integration. This makes them an ideal target for genetic transformation, enabling the introduction of traits such as enhanced disease resistance, improved yield, and increased tolerance to abiotic stresses. In some plants such as cotton genetic transformation via SE is the most effective and commonly used approach due to the recalcitrance of these species to other transformation methods (Firoozabady & DeBoer, 1993). Beyond its practical applications, SE is also invaluable for the conservation of plant germplasm, particularly for rare or endangered species (Xiong et al., 2022). By enabling the *in vitro* preservation and multiplication of plant genetic resources, SE helps safeguard biodiversity and ensures the availability of valuable genetic material for future breeding and research efforts. Furthermore, SE serves as an excellent model system for studying the cellular, molecular, and genetic mechanisms underlying embryo development (Karami et al., 2009). Research in this area provides crucial insights into the processes of cell differentiation, pattern formation, and morphogenesis, contributing to our broader understanding of plant development and growth. These studies not only enhance our knowledge of plant biology but also inform the development of more efficient and effective strategies for plant propagation and improvement.

4.8 CONCLUSION

SE stands as a crucial pathway in plant regeneration, underscoring the remarkable totipotency inherent in plant cells. This process, wherein differentiated somatic cells are reprogrammed to form structures analogous to zygotic embryos, has profound implications for both fundamental plant biology and applied sciences such as agriculture and biotechnology. Central to the success of SE are the molecular factors that orchestrate this complex reprogramming. Plant growth regulators, particularly auxins and cytokinins, play a pivotal role in initiating and sustaining the embryogenic potential of somatic cells. Additionally, specific transcription factors and signaling molecules guide the cells through various stages of embryogenesis (Table 4.1), ensuring the proper expression of genes necessary for embryo formation and development. These molecular insights not only enhance our understanding of plant developmental processes but also provide powerful tools for manipulating embryogenesis in diverse plant species. The practical applications of SE are far-reaching. It offers an efficient means for the clonal propagation of elite plant varieties, the conservation of endangered species, and the production of uniform, disease-free plants on a large scale. Moreover, SE serves as a critical platform for plant genetic engineering, facilitating the introduction of new traits with precision and enabling the development of crops that are better suited to meet global agricultural challenges. Looking forward, the integration of SE with cutting-edge technologies, such as CRISPR-based gene editing, promises to advance our ability to engineer plants with desirable characteristics more efficiently. As our understanding of the molecular underpinnings of this process deepens, we can expect further refinements in the techniques and broader applications in both research and industry. Therefore, SE is not merely a pathway of regeneration; it is a fundamental aspect of plant biotechnology with significant implications for the future of agriculture and environmental sustainability. The continued exploration of its molecular basis and the innovative application of this knowledge will ensure that SE remains at the forefront of plant science, offering solutions to some of the most pressing challenges facing humanity.

REFERENCES

Altamura, M. M., Rovere, F. D. Fattorini, L., D'Angeli, S., & Falasca, G. (2016, January 1). Recent advances on genetic and physiological bases of in vitro somatic embryo formation. *Methods in Molecular Biology*, 1359, 47–85. https://doi.org/10.1007/978-1-4939-3061-6_3

Asghar, S., Ghori, N., Hyat, F., Li, Y., & Chen, C. (2022, November 9). Use of auxin and cytokinin for somatic embryogenesis in plant: A story from competence towards completion. *Plant Growth Regulation*, 99(3), 413–428. https://doi.org/10.1007/s10725-022-00923-9

Avilez-Montalvo, J. R., Quintana-Escobar, A. O., Méndez-Hernández, H. A., Aguilar-Hernández, V., Brito-Argáez, L., Galáz-Ávalos, R. M., Uc-Chuc, M. A., & Loyola-Vargas, V. M. (2022, August 2). Auxin-cytokinin cross talk in somatic embryogenesis of coffea canephora. *Plants*, 11(15), 2013. https://doi.org/10.3390/plants11152013

Babak, D. H., Azimi, A. A., Majd, A., & Hassan, E. (2008, June 3). Abnormal plantlets regeneration through direct somatic embryogenesis on immature seeds of *Vinca herbacea Waldst.* and Kit. *African Journal of Biotechnology*, 7(11), 1679–1683. https://doi.org/10.5897/ajb08.111

Bidabadi, S. S., & Jain, S. M. (2020, June 1). Cellular, molecular, and physiological aspects of in vitro plant regeneration. *Plants*, 9(6), 702. https://doi.org/10.3390/plants9060702

Bornman, C. H. (1993, January 1). Micropropagation and somatic embryogenesis. In *Plant Breeding: Principles and Prospects* (pp. 246–260). Dordrecht: Springer Netherlands. https://doi.org/10.1007/978-94-011-1524-7_17

Boutilier, K., Offringa, R., Sharma, V. K., Kieft, H., Ouellet, T., Zhang, L., Hattori, J., Liu, C., Lammeren, A. V., Miki, B., Custers, J., & Campagne, M. M. V. L. (2002, July 18). Ectopic expression of BABY BOOM triggers a conversion from vegetative to embryonic growth. *The Plant Cell*, 14(8), 1737–1749. https://doi.org/10.1105/tpc.001941

Braybrook, S. A., Stone, S. L., Park, S., Bui, A., Le, B. H., Fischer, R. L., Goldberg, R. B., & Harada, J. J. (2006, February 21). Genes directly regulated by LEAFY COTYLEDON2 provide insight into the control of embryo maturation and somatic embryogenesis. *Proceedings of the National Academy of Sciences of the United States of America*, 103(9), 3468–3473. https://doi.org/10.1073/pnas.0511331103

Cao, Y. (2013, January 1). Regulation of germ layer formation by pluripotency factors during embryogenesis. *Cell & Bioscience*, 3(1), 15. https://doi.org/10.1186/2045-3701-3-15

Carneros, E., Sánchez-Muñoz, J., Pérez-Pérez, Y., Pintos, B., Garay, A. G., & Testillano, P. S. (2023, April 3). Dynamics of endogenous auxin and its role in somatic embryogenesis induction and progression in cork oak. *Plants*, 12(7), 1542. https://doi.org/10.3390/plants12071542

Chakrabarty, D., Yu, K., & Paek, K. (2003, July 1). Detection of DNA methylation changes during somatic embryogenesis of Siberian ginseng (*Eleuterococcus senticosus*). *Plant Science*, 165(1), 61–68. https://doi.org/10.1016/s0168-9452(03)00127-4

Chen, B., Maas, L., Figueiredo, D. D., Zhong, Y., Reis, R. S., Li, M., Horstman, A., Riksen, T., Weemen, M., Liu, H., Siemons, C., Chen, S., Angenent, G. C., & Boutilier, K. (2022, June 16). BABY BOOM regulates early embryo and endosperm development. *Proceedings of the National Academy of Sciences of the United States of America*, 119(25). https://doi.org/10.1073/pnas.2201761119

Chen, S., Kurdyukov, S., Kereszt, A., Wang, X., Gresshoff, P. M., & Rose, R. J. (2009, July 29). The association of homeobox gene expression with stem cell formation and morphogenesis in cultured *Medicago truncatula*. *Planta*, 230(4), 827–840. https://doi.org/10.1007/s00425-009-0988-1

Dodeman, V. L., Ducreux, G., & Kreis, M. (1997, January 1). REVIEW ARTICLE: Zygotic embryogenesis versus somatic embryogenesis. *Journal of Experimental Botany*, 48(8), 1493–1509. https://doi.org/10.1093/jxb/48.8.1493

Dunwell, J. M. (2010, May 1). Haploids in flowering plants: Origins and exploitation. *Plant Biotechnology Journal*, 8(4), 377–424. https://doi.org/10.1111/j.1467-7652.2009.00498.x

Fambrini, M., Usai, G., & Pugliesi, C. (2022, December 15). Induction of somatic embryogenesis in plants: Different players and focus on WUSCHEL and WUS-RELATED HOMEOBOX (WOX) transcription factors. *International Journal of Molecular Sciences*, 23(24), 15950. https://doi.org/10.3390/ijms232415950

Fehér, A. (2015, April 1). Somatic embryogenesis — Stress-induced remodeling of plant cell fate. *Biochimica et Biophysica Acta (BBA)-Gene Regulatory Mechanisms*, 1849(4), 385–402. https://doi.org/10.1016/j.bbagrm.2014.07.005

Fehér, A. (2019, April 26). Callus, dedifferentiation, totipotency, somatic embryogenesis: What these terms mean in the era of molecular plant biology? *Frontiers in Plant Science*, 10. https://doi.org/10.3389/fpls.2019.00536

Firoozabady, E., & DeBoer, D. L. (1993, October 1). Plant regeneration via somatic embryogenesis in many cultivars of cotton (*Gossypium hirsutum* L.). *In Vitro Cellular & Developmental Biology-Plant*, 29(4), 166–173. https://doi.org/10.1007/bf02632030

Frost, J. M., Rhee, J. H., & Choi, Y. (2024, October 1). Dynamics of DNA methylation and its impact on plant embryogenesis. *Current Opinion in Plant Biology*, 81, 102593. https://doi.org/10.1016/j.pbi.2024.102593

Gallois, J., Nora, F. R., Mizukami, Y., & Sablowski, R. (2004, February 15). WUSCHEL induces shoot stem cell activity and developmental plasticity in the root meristem. *Genes & Development*, 18(4), 375–380. https://doi.org/10.1101/gad.291204

Gambino, G., Minuto, M., Boccacci, P., Perrone, I., Vallania, R., & Gribaudo, I. (2010, December 1). Characterization of expression dynamics of WOX homeodomain transcription factors during somatic embryogenesis in *Vitis vinifera*. *Journal of Experimental Botany,* 62(3), 1089–1101. https://doi.org/10.1093/jxb/erq349

Goldberg, R. B., Paiva, G. D., & Yadegari, R. (1994, October 28). Plant embryogenesis: Zygote to seed. *Science*, 266(5185), 605–614. https://doi.org/10.1126/science.266.5185.605

Graaff, E. V. D., Laux, T., & Rensing, S. A. (2009, January 1). The WUS homeobox-containing (WOX) protein family. *Genome Biology*, 10(12), 248. https://doi.org/10.1186/gb-2009-10-12-248

Gray, D. J., Purohit, A., & Triglano, R. N. (1991, January 1). Somatic embryogenesis and development of synthetic seed technology. *Critical Reviews in Plant Sciences*, 10(1), 33–61. https://doi.org/10.1080/07352689109382306

Guilfoyle, T. J., & Hagen, G. (2007, October 1). Auxin response factors. *Current Opinion in Plant Biology*, 10(5), 453–460. https://doi.org/10.1016/j.pbi.2007.08.014

Gulzar, B., Mujib, A., Malik, M. Q., Sayeed, R., Mamgain, J., & Ejaz, B. (2020, December 1). Genes, proteins and other networks regulating somatic embryogenesis in plants. *Journal of Genetic Engineering and Biotechnology*, 18(1), 31. https://doi.org/10.1186/s43141-020-00047-5

Guo, F., Liu, C., Xia, H., Bi, Y., Zhao, C., Zhao, S., Hou, L., Li, F., & Wang, X. (2013, August 12). Induced expression of *AtLEC1* and *AtLEC2* differentially promotes somatic embryogenesis in transgenic tobacco plants. *PLoS One*, 8(8), e71714. https://doi.org/10.1371/journal.pone.0071714

Harada, J. J., Belmonte, M. F., & Kwong, R. W. M. (2010, October 18). *Plant Embryogenesis (Zygotic and Somatic)*. https://doi.org/10.1002/9780470015902.a0002042.pub2

Harada, J. J., Lotan, T., Fischer, R. L., & Goldberg, R. B. (1998, December 1)....response: Embryos without sex. *Trends in Plant Science*, 3(12), 452–453. https://doi.org/10.1016/s1360-1385(98)01347-8

Heck, G. R., Perry, S. E., Nichols, K. W., & Fernandez, D. E. (1995, August 1). AGL15, a MADS domain protein expressed in developing embryos. *The Plant Cell*, 7(8), 1271–1282. https://doi.org/10.1105/tpc.7.8.1271

Horstman, A., Bemer, M., & Boutilier, K. (2017, August 1). A transcriptional view on somatic embryogenesis. *Regeneration*, 4(4), 201–216. https://doi.org/10.1002/reg2.91

Ikeda, M., & Ohme-Takagi, M. (2014, September 3). TCPs, WUSs, and WINDs: Families of transcription factors that regulate shoot meristem formation, stem cell maintenance, and somatic cell differentiation. *Frontiers in Plant Science*, 5. https://doi.org/10.3389/fpls.2014.00427

Iwase, A., Harashima, H., Ikeuchi, M., Rymen, B., Ohnuma, M., Komaki, S., Morohashi, K., Kurata, T., Nakata, M., Ohme-Takagi, M., Grotewold, E., & Sugimoto, K. (2016, December 23). WIND1 promotes shoot regeneration through transcriptional activation of *ENHANCER OF SHOOT REGENERATION1* in Arabidopsis. *The Plant Cell*, 29(1), 54–69. https://doi.org/10.1105/tpc.16.00623

Iwase, A., Kondo, Y., Laohavisit, A., Takebayashi, A., Ikeuchi, M., Matsuoka, K., Asahina, M., Mitsuda, N., Shirasu, K., Fukuda, H., & Sugimoto, K. (2021, August 10). WIND transcription factors orchestrate wound-induced callus formation, vascular reconnection and defense response in Arabidopsis. *The New Phytologist*, 232(2), 734–752. https://doi.org/10.1111/nph.17594

Iwase, A., Mita, K., Favero, D. S., Mitsuda, N., Sasaki, R., Kobayshi, M., Takebayashi, Y., Kojima, M., Kusano, M., Oikawa, A., Sakakibara, H., Saito, K., Imamura, J., & Sugimoto, K. (2018, October 1). WIND1 induces dynamic metabolomic reprogramming during regeneration in *Brassica napus*. *Developmental Biology*, 442(1), 40–52. https://doi.org/10.1016/j.ydbio.2018.07.006

Iwase, A., Mita, K., Nonaka, S., Ikeuchi, M., Koizuka, C., Ohnuma, M., Ezura, H., Imamura, J., & Sugimoto, K. (2015, March 26). WIND1-based acquisition of regeneration competency in Arabidopsis and rapeseed. *Journal of Plant Research*, 128(3), 389–397. https://doi.org/10.1007/s10265-015-0714-y

Iwase, A., Mitsuda, N., Koyama, T., Hiratsu, K., Kojima, M., Arai, T., Inoue, Y., Seki, M., Sakakibara, H., Sugimoto, K., & Ohme-Takagi, M. (2011, March 1). The AP2/ERF transcription factor WIND1 controls cell dedifferentiation in Arabidopsis. *Current Biology*, 21(6), 508–514. https://doi.org/10.1016/j.cub.2011.02.020

Ji, A., Geng, X., Zhang, Y., & Guo-liang, W. (2011, January 1). Advances in somatic embryogenesis research of horticultural plants. *American Journal of Plant Sciences*, 2(6), 727–732. https://doi.org/10.4236/ajps.2011.26087

Joshi, S., Paul, P., Hartman, J. M., & Perry, S. E. (2022, March 28). AGL15 promotion of somatic embryogenesis: Role and molecular mechanism. *Frontiers in Plant Science*, 13. https://doi.org/10.3389/fpls.2022.861556

Junker, A., & Bäumlein, H. (2012, December 1). Multifunctionality of the LEC1 transcription factor during plant development. *Plant Signaling & Behavior*, 7(12), 1718–1720. https://doi.org/10.4161/psb.22365

Karami, O., Aghavaisi, B., & Pour, A. M. (2009, September 10). Molecular aspects of somatic-to-embryogenic transition in plants. *Journal of Chemical Biology*, 2(4), 177–190. https://doi.org/10.1007/s12154-009-0028-4

Karami, O., Philipsen, C., Rahimi, A., Nurillah, A. R., Boutilier, K., & Offringa, R. (2022, November 28). Endogenous auxin maintains embryonic cell identity and promotes somatic embryo development in Arabidopsis. *The Plant Journal*, 113(1), 7–22. https://doi.org/10.1111/tpj.16024

Khai, H. D., Bien, L. T., Vịnh, N. Q., Đoàn, M. D., Nghiep, N. D., Nhu, N. T., Tùng, H. T., Luận, V. Q., Cường, Đ. M., & Nhựt, D. T. (2021, November 1). Alterations in endogenous hormone levels and energy metabolism promoted the induction, differentiation and maturation of Begonia somatic embryos under clinorotation. *Plant Science*, 312, 111045. https://doi.org/10.1016/j.plantsci.2021.111045

Kulińska-Łukaszek, K., Tobojka, M., Adamiok-Ostrowska, A., & Kurczyńska, E. (2012, June 1). Expression of the *BBM* gene during somatic embryogenesis of *Arabidopsis thaliana*. *Biologia Plantarum*, 56(2), 389–394. https://doi.org/10.1007/s10535-012-0105-3

Ledwoń, A., & Gaj, M. D. (2011, March 29). *LEAFY COTYLEDON1*, *FUSCA3* expression and auxin treatment in relation to somatic embryogenesis induction in Arabidopsis. *Plant Growth Regulation*, 65(1), 157–167. https://doi.org/10.1007/s10725-011-9585-y

Lee, H., Fischer, R. L., Goldberg, R. B., & Harada, J. J. (2003, February 18). Arabidopsis LEAFY COTYLEDON1 represents a functionally specialized subunit of the CCAAT binding transcription factor. *Proceedings of the National Academy of Sciences of the United States of America*. https://www.pnas.org/doi/full/10.1073/pnas.0437909100

Li, J., Zhang, Q., Wang, Z., & Liu, Q. (2024, January 30). The roles of epigenetic regulators in plant regeneration: Exploring patterns amidst complex conditions. *Plant Physiology*, 194(4), 2022–2038. https://doi.org/10.1093/plphys/kiae042

Li, L. (2023, November 10). *In vitro* somatic embryogenesis in turf-type bermudagrass: Roles of abscisic acid and gibberellic acid, and occurrence of secondary somatic embryogenesis. *Plant Breeding*. https://onlinelibrary.wiley.com/doi/10.1046/j.1439-0523.2002.00684.x

Li, S., Xie, Z., Hu, C., & Zhang, J. (2016, February 3). A review of auxin response factors (ARFs) in plants. *Frontiers in Plant Science*, 7. https://doi.org/10.3389/fpls.2016.00047

Liang, Z., Yuan, L., Xiong, X., Hao, Y., Song, X., Zhu, T., Yu, Y., Fu, W., Lei, Y., Xu, J., Liu, J., Li, J., & Li, C. (2022, July 22). The transcriptional repressors VAL1 and VAL2 mediate genome-wide recruitment of the CHD3 chromatin remodeler PICKLE in Arabidopsis. *The Plant Cell*, 34(10), 3915–3935. https://doi.org/10.1093/plcell/koac217

Liu, B., Sun, G., Liu, C., & Liu, S. (2021, November 26). LEAFY COTYLEDON 2: A regulatory factor of plant growth and seed development. *Genes*, 12(12), 1896. https://doi.org/10.3390/genes12121896

Liu, Z., Ge, X., Qiu, W., Long, J., Jia, H., Yang, W., Dutt, M., Wu, X., & Guo, W. (2018, December 1). Overexpression of the *CsFUS3* gene encoding a B3 transcription factor promotes somatic embryogenesis in *Citrus*. *Plant Science*, 277, 121–131. https://doi.org/10.1016/j.plantsci.2018.10.015

Luerssen, H., Kirik, V., Herrmann, P., & Miséra, S. (1998, September 1). FUSCA3encodes a protein with a conserved VP1/ABI3-like B3 domain which is of functional importance for the regulation of seed maturation in *Arabidopsis thaliana*. *The Plant Journal*, 15(6), 755–764. https://doi.org/10.1046/j.1365-313x.1998.00259.x

Luo, J., Zhou, J., & Zhang, J. (2018, January 16). *Aux/IAA* gene family in plants: Molecular structure, regulation, and function. *International Journal of Molecular Sciences*, 19(1), 259. https://doi.org/10.3390/ijms19010259

Magnani, E., Jiménez-Gómez, J. M., Soubigou-Taconnat, L., Lepiniec, L., & Fiume, E. (2017, December 1). Profiling the onset of somatic embryogenesis in Arabidopsis. *BMC Genomics*, 18(1). https://doi.org/10.1186/s12864-017-4391-1

McKersie, B. D., & Brown, D. C. (1996, September 1). Somatic embryogenesis and artificial seeds in forage legumes. *Seed Science Research*, 6(3). 109–126

Méndez-Hernández, H. A., Ledezma-Rodríguez, M., Avilez-Montalvo, R. N., Juárez-Gómez, Y. L., Skeete, A. O., Avilez-Montalvo, J. R., De-la-Peña, C., & Loyola-Vargas, V. M. (2019, February 7). Signaling overview of plant somatic embryogenesis. *Frontiers in Plant Science*, 10. https://doi.org/10.3389/fpls.2019.00077

Merkle, S. A., Montello, P., Xiu-qin, X., Upchurch, B. L., & Smith, D. R. (2006, February 1). Light quality treatments enhance somatic seedling production in three southern pine species. *Tree Physiology*, 26(2), 187–194. https://doi.org/10.1093/treephys/26.2.187

Meucci, A., Ghelardi, C., Maggini, R., Malorgio, F., Chietera, G., & Mensuali-Sodi, A. (2024, July 23). Micropropagation via somatic embryogenesis of *Iris pallida* Lam. ecotypes. *Plant Cell, Tissue and Organ Culture (PCTOC)*, 158(2). https://doi.org/10.1007/s11240-024-02818-1

Miguel, C. M. G., & Marum, L. (2011, May 26). An epigenetic view of plant cells cultured in vitro: Somaclonal variation and beyond. *Journal of Experimental Botany*, 62(11), 3713–3725. https://doi.org/10.1093/jxb/err155

Mira, M. M., Elshanshory, A., & Stasolla, C. (2015, June 1). Ethylene is integrated into the nitric oxide regulation of Arabidopsis somatic embryogenesis. *Journal of Genetic Engineering and Biotechnology*, 13(1), 7–17. https://doi.org/10.1016/j.jgeb.2015.01.001

Möller, B., & Weijers, D. (2009, September 30). Auxin control of embryo patterning. *Cold Spring Harbor Perspectives in Biology*, 1(5), a001545. https://doi.org/10.1101/cshperspect.a001545

Moore, L., Le, T. T., & Fan, G. (2012, July 11). DNA methylation and its basic function. *Neuropsychopharmacology*, 38(1), 23–38. https://doi.org/10.1038/npp.2012.112

Nambara, E., Hayama, R., Tsuchiya, Y., Nishimura, M., Kawaide, H., Kamiya, Y., & Naito, S. (2000, April 1). The role of *ABI3* and *FUS3* loci in *Arabidopsis thaliana* on phase transition from late embryo development to germination. *Developmental Biology*, 220(2), 412–423. https://doi.org/10.1006/dbio.2000.9632

Nolan, K. E., & Rose, R. J. (2010, March 5). *Plant Regeneration – Somatic Embryogenesis*, pp. 39–59. https://doi.org/10.1002/9780470686522.ch3

Nowak, K., Morończyk, J., Wójcik, A. M., & Gaj, M. D. (2020, September 14). AGL15 controls the embryogenic reprogramming of somatic cells in Arabidopsis through the histone acetylation-mediated repression of the miRNA biogenesis genes. *International Journal of Molecular Sciences*, 21(18), 6733. https://doi.org/10.3390/ijms21186733

Ochatt, S., & Revilla, M. A. (2016, January 1). From stress to embryos: Some of the problems for induction and maturation of somatic embryos. In *In Vitro Embryogenesis in Higher Plants*, pp. 523–536. https://doi.org/10.1007/978-1-4939-3061-6_31

Ogas, J., Cheng, J., Sung, Z. R., & Somerville, C. (1997, July 4). Cellular differentiation regulated by gibberellin in the *Arabidopsis thaliana pickle* mutant. *Science*, 277(5322), 91–94. https://doi.org/10.1126/science.277.5322.91

Pandey, D. K., & Chaudhary, B. (2014a, January 1). Oxidative stress responsive SERK1 gene directs the progression of somatic embryogenesis in cotton (*Gossypium hirsutum* L. cv. Coker 310). *American Journal of Plant Sciences*, 5, 80–102. https://doi.org/10.4236/ajps.2014.51012

Pandey, D. K., & Chaudhary, B. (2014b, January 1). Role of plant somatic embryogenesis receptor kinases (SERKs) in cell-to-embryo transitional activity: Key at novel assorted structural subunits. *American Journal of Plant Sciences*, 5(21), 3177–3193. https://doi.org/10.4236/ajps.2014.521334

Pandey, D. K., & Chaudhary, B. (2015, January 1). Genes and *trans*-factors underlying embryogenic transition in plant soma-cells. In *Advances in the Understanding of Biological Sciences Using Next Generation Sequencing (NGS) Approaches*, pp. 155–178. https://doi.org/10.1007/978-3-319-17157-9_10

Pandey, D. K., Singh, A. K., & Chaudhary, B. (2012, October 2). Boron-mediated plant somatic embryogenesis: A provocative model. *Journal of Botany*, 2012, 375829. https://doi.org/10.1155/2012/375829

Ramakrishnan, M., Zhou, M., Ceasar, S. A., Ali, D. J., Maharajan, T., Vinod, K. K., Sharma, A., Ahmad, Z., & Wei, Q. (2023, October 4). Epigenetic modifications and miRNAs determine the transition of somatic cells into somatic embryos. *Plant Cell Reports*, 42(12), 1845–1873. https://doi.org/10.1007/s00299-023-03071-0

Rihan, H. Z., Kareem, F., El-Mahrouk, M. E., & Fuller, M. P. (2017, November 3). Artificial seeds (principle, aspects and applications). *Agronomy*, 7(4), 71. https://doi.org/10.3390/agronomy7040071

Rocha, D. I., & Dornelas, M. C. (2013, September 18). *Molecular Overview on Plant Somatic Embryogenesis*, pp. 1–17. https://doi.org/10.1079/pavsnnr20138022

Salaün, C., Lepiniec, L., & Dubreucq, B. (2021, July 17). Genetic and molecular control of somatic embryogenesis. *Plants*, 10(7), 1467. https://doi.org/10.3390/plants10071467

Shen, K., Qu, M., & Zhao, P. (2023, January 5). The roads to haploid embryogenesis. *Plants*, 12(2), 243. https://doi.org/10.3390/plants12020243

Siddiqui, Z. H., Abbas, Z. K., Ansari, M. W., & Khan, M. N. (2019, September 1). The role of miRNA in somatic embryogenesis. *Genomics*, 111(5), 1026–1033. https://doi.org/10.1016/j.ygeno.2018.11.022

Smet, I. D., Lau, S., Mayer, U., & Jürgens, G. (2010, March 1). Embryogenesis – the humble beginnings of plant life. *The Plant Journal*, 61(6), 959–970. https://doi.org/10.1111/j.1365-313x.2010.04143.x

Stone, S. L., Braybrook, S. A., Paula, S. L., Kwong, L. W., Meuser, J., Pelletier, J., Hsieh, T., Fischer, R. L., Goldberg, R. B., & Harada, J. J. (2023a, December 4). *Arabidopsis* LEAFY COTYLEDON2 induces maturation traits and auxin activity: Implications for somatic embryogenesis. *Proceedings of the National Academy of Sciences of the United States of America*. https://www.pnas.org/doi/full/10.1073/pnas.0712364105

Stone, S. L., Kwong, L. W., Matsudaira, K., Pelletier, J., Lepiniec, L., Fischer, R. L., Goldberg, R. B., & Harada, J. J. (2023b, November 10). *LEAFY COTYLEDON2* encodes a B3 domain transcription factor that induces embryo development. *Proceedings of the National Academy of Sciences of the United States of America*. https://pnas.org/doi/full/10.1073/pnas.201413498

Su, Y. H., & Zhang, X. S. (2009, July 1). Auxin gradients trigger de novo formation of stem cells during somatic embryogenesis. *Plant Signaling & Behavior*, 4(7), 574–576. https://doi.org/10.4161/psb.4.7.8730

Su, Y. H., Zhao, X., Liu, Y. B., Zhang, C. L., O'Neill, S. D., & Zhang, X. S. (2009, July 24). Auxin-induced *WUS* expression is essential for embryonic stem cell renewal during somatic embryogenesis in Arabidopsis. *The Plant Journal*, 59(3), 448–460. https://doi.org/10.1111/j.1365-313x.2009.03880.x

Tang, L. P., Zhang, X. S., & Su, Y. H. (2020, July 1). Regulation of cell reprogramming by auxin during somatic embryogenesis. *Abiotech*, 1(3), 185–193. https://doi.org/10.1007/s42994-020-00029-8

Tapingkae, T., Zulkarnain, Z., Kawaguchi, M., Ikeda, T., & Taji, A. (2012, January 1). Somatic (asexual) procedures (haploids, protoplasts, cell selection) and their applications. In *Plant Biotechnology and Agriculture*, pp. 141–162. https://doi.org/10.1016/b978-0-12-381466-1.00010-9

Thakare, D., Tang, W., Hill, K., & Perry, S. E. (2008, February 27). The MADS-domain transcriptional regulator AGAMOUS-LIKE15 promotes somatic embryo development in Arabidopsis and soybean. *Plant Physiology*, 146(4), 1663–1672. https://doi.org/10.1104/pp.108.115832

Theimer, R. R., Kudielka, R. A., & Rösch, I. (1986, July 1). Induction of somatic embryogenesis in anise in microgravity. *Naturwissenschaften*, 73(7), 442–443. https://doi.org/10.1007/bf00367292

Tian, L., & Brown, D. C. (2000, April 1). Improvement of soybean somatic embryo development and maturation by abscisic acid treatment. *Canadian Journal of Plant Science*, 80(2), 271–276. https://doi.org/10.4141/p99-056

Tisserat, B., Esan, E. B., & Murashige, T. (1979, January 1). *Somatic Embryogenesis in Angiosperms*, pp. 1–78. https://doi.org/10.1002/9781118060742.ch1

Tsuchiya, Y., Nambara, E., Naito, S., & Mccourt, P. (2023, November 10). The *FUS3* transcription factor functions through the epidermal regulator *TTG1* during embryogenesis in *Arabidopsis*. https://onlinelibrary.wiley.com/doi/10.1046/j.1365-313X.2003.01939.x

Vogel, G. (2005). How does a single somatic cell become a whole plant? *Science*, 309, 86. https://doi.org/10.1126/science.309.5731.86

Vishal, B., & Kumar, P. P. (2018, June 20). Regulation of seed germination and abiotic stresses by gibberellins and abscisic acid. *Frontiers in Plant Science*, 9. https://doi.org/10.3389/fpls.2018.00838

Wang, F., & Perry, S. E. (2013, January 11). Identification of direct targets of FUSCA3, a key regulator of arabidopsis seed development. *Plant Physiology*, 161(3), 1251–1264. https://doi.org/10.1104/pp.112.212282

Wang, H., Caruso, L. V., Downie, A. B., & Perry, S. E. (2004, April 14). The embryo MADS domain protein AGAMOUS-Like 15 directly regulates expression of a gene encoding an enzyme involved in gibberellin metabolism. *The Plant Cell*, 16(5), 1206–1219. https://doi.org/10.1105/tpc.021261

Wójcik, A. M., Wójcikowska, B., & Gaj, M. D. (2020, February 16). Current perspectives on the auxin-mediated genetic network that controls the induction of somatic embryogenesis in plants. *International Journal of Molecular Sciences*, 21(4), 1333. https://doi.org/10.3390/ijms21041333

Wójcikowska, B., & Gaj, M. D. (2017, March 2). Expression profiling of *AUXIN RESPONSE FACTOR* genes during somatic embryogenesis induction in *Arabidopsis*. *Plant Cell Reports*, 36(6), 843–858. https://doi.org/10.1007/s00299-017-2114-3

Wójcikowska, B., Jaskóła, K., Gąsiorek, P., Meus, M., Nowak, K., & Gaj, M. D. (2013, May 31). *LEAFY COTYLEDON2* (*LEC2*) promotes embryogenic induction in somatic tissues of Arabidopsis, via *YUCCA*-mediated auxin biosynthesis. *Planta*, 238(3), 425–440. https://doi.org/10.1007/s00425-013-1892-2

Wójcikowska, B., Wójcik, A. M., & Gaj, M. D. (2020, March 26). Epigenetic regulation of auxin-induced somatic embryogenesis in plants. *International Journal of Molecular Sciences*, 21(7), 2307. https://doi.org/10.3390/ijms21072307

Wu, X., Chory, J., & Weigel, D. (2007, September 1). Combinations of WOX activities regulate tissue proliferation during Arabidopsis embryonic development. *Developmental Biology*, 309(2), 306–316. https://doi.org/10.1016/j.ydbio.2007.07.019

Xiong, Y., Chen, S., Wu, T., Wu, K., Li, Y., Zhang, X., Silva, J. A. T. D., Zeng, S., & Ma, G. (2022, November 28). Shoot organogenesis and somatic embryogenesis from leaf and petiole explants of endangered *Euryodendron excelsum*. *Scientific Reports*, 12(1). https://doi.org/10.1038/s41598-022-24744-y

Xu, L., & Huang, H. (2014, January 1). Genetic and epigenetic controls of plant regeneration. *Current Topics in Developmental Biology*, 108, 1–33. https://doi.org/10.1016/b978-0-12-391498-9.00009-7

Zhang, H., Bishop, B., Ringenberg, W., Muir, W. M., & Ogas, J. (2012, March 27). The CHD3 remodeler PICKLE associates with genes enriched for trimethylation of histone H3 lysine 27. *Plant Physiology*, 159(1), 418–432. https://doi.org/10.1104/pp.112.194878

Zheng, Q., Zheng, Y., & Perry, S. E. (2013, March 1). AGAMOUS-Like15 promotes somatic embryogenesis in Arabidopsis and soybean in part by the control of ethylene biosynthesis and response. *Plant Physiology*, 161(4), 2113–2127. https://doi.org/10.1104/pp.113.216275

Zheng, Q., Zheng, Y., Ji, H., Burnie, W., & Perry, S. E. (2016, October 28). Gene regulation by the AGL15 transcription factor reveals hormone interactions in somatic embryogenesis. *Plant Physiology*, 172(4), 2374–2387. https://doi.org/10.1104/pp.16.00564

Zhou, J., & Luo, J. (2018, September 14). The PIN-FORMED auxin efflux carriers in plants. *International Journal of Molecular Sciences*, 19(9), 2759. https://doi.org/10.3390/ijms19092759

Zhou, X., Guo, Y., Zhao, P., & Sun, M. (2018, March 8). Comparative analysis of WUSCHEL-related homeobox genes revealed their parent-of-origin and cell type-specific expression pattern during early embryogenesis in tobacco. *Frontiers in Plant Science*, 9. https://doi.org/10.3389/fpls.2018.00311

Zimmerman, J. L. (1993, October 1). Somatic embryogenesis: A model for early development in higher plants. *The Plant Cell*, 5(10), 1411–1423. https://doi.org/10.1105/tpc.5.10.1411

Zuo, J., Niu, Q., Frugis, G., & Chua, N. (2002, May 1). The *WUSCHEL* gene promotes vegetative-to-embryonic transition in *Arabidopsis*. *The Plant Journal*, 30(3), 349–359. https://doi.org/10.1046/j.1365-313x.2002.01289.x

5 Unlocking the Potential of miRNAs for Plant *In Vitro* Regeneration

Sakshi Arora, Priyasha Das, and Bhupendra Chaudhary

5.1 INTRODUCTION

Plants possess totipotency and pluripotency abilities and, therefore, display a high degree of development plasticity and regeneration ability (Ikeuchi et al. 2016). Plants can regenerate from adult tissues, organs, calli, and even single cells under *in vitro* conditions. This technology has revolutionized agriculture and plant biotechnology by enabling rapid propagation, genetic manipulation, and the development of new varieties (Ikeuchi et al. 2016; Long et al. 2022). *In vitro* plant regeneration involves the manipulation of the genetic control of the cultured cell by providing a nutrient-rich environment with appropriate phytohormones and growth factors. Thus, the success of a particular plant tissue culture technology consists of the influence of multiple intrinsic and extrinsic factors, including plant genotype, explant type and age, phytohormone, and the medium used (Ikeuchi et al. 2016).

Organogenesis and somatic embryogenesis (SE) are two major modes employed for *in vitro* plant regeneration, and both can proceed directly from explant, or indirectly by generating calli (Bidabadi and Jain 2020). Organogenesis encompasses the direct development of new organs, either adventitious or *de novo*, from explant tissues, whereas SE entails initiation by forming structures resembling embryos from somatic cells (Duclercq et al. 2011; Zimmerman 1993). The regeneration process generally involves several steps involving rapid cell division and (de)differentiation. It is tightly regulated through a complex cellular mechanism in response to phytohormone signaling and the external environment. The complex mechanism includes the activation of various genes, including meristem-specific genes, transcription factors, signaling pathways, and epigenetic operations, including DNA methylation, histone modification, and small-RNA-mediated gene expression regulation (Hill and Schaller 2013; Ikeuchi et al. 2016; Su and Zhang 2014).

Small RNAs are the product of a noncoding RNA gene, resemble protein-coding genes, and have a length typically less than 200 nucleotides (usually ranging from 20 to 30) (Zhan and Meyers 2023). There are various kinds of small RNAs, including miRNAs and short-interfering RNAs (siRNAs), which are characterized in plants and are reported to regulate gene expression through chromatin modification, posttranscription gene silencing, or translation inhibition (Zhan and Meyers 2023). miRNAs

DOI: 10.1201/9781032702391-5

are 21–24 nucleotide long small RNA, transcribed from the noncoding genic region, including the intronic region by DNA polymerase II, and are processed from the primary-miRNA (*pri*-miRNA) through two or more RNAase III-mediated steps (Chen et al. 2020; Jung et al. 2009). The first step involves the conversion of *pri*-miRNA by the *DCL1* enzyme to the precursor (pre)-miRNA, which is a stem-loop structure. The second cleavage occurs on *pre*-miRNA in the nucleus by the same enzyme and leads to the formation of miRNA:miRNA duplex. The miRNA duplex is then transported to the cytoplasm through the HASTY protein, which is transformed into single-stranded mature miRNA that subsequently binds with the RISC complex (Bartel 2004; Jung et al. 2009; Chen et al. 2020). The mature miRNAs are evolutionary conserved, exhibit complementation with target gene mRNA, and modulate gene expression primarily by cleaving mRNA through the RISC complex at the posttranscriptional level. In some cases, they might obstruct the process of converting genetic information into proteins or, in exceptional situations, trigger changes in the DNA structure that affect gene expression (Baumberger and Baulcombe 2005; Chen 2005).

miRNAs generally target transcription factors that modulate the expression of various target genes and are reported to have significant contributions in all plant growth and development processes, including seed germination, plant organ development, and inflorescence development. Moreover, miRNAs are responsive to various intrinsic and extensive factors, including temperature, day length, nutrient starvation, hormonal signaling, and biotic stress, and thus are considered master regulators of gene expression (Shriram et al. 2016; Sunkar et al. 2012). This chapter discusses the role of miRNAs during plant regeneration, including organogenesis and SE.

5.2 mIRNA REGULATION OF *IN VITRO* ORGANOGENESIS

Organogenesis is an *in vitro* technique of plant regeneration where plant cells from the preexisting organs or tissues differentiate into meristematic cells that further develop adventitious shoots or roots (Duclercq et al. 2011; Lardon and Geelen 2020). Organogenesis is direct if root and shoot organs are induced directly from preexisting organs/tissue. However, if a preexisting tissue/organ first generates a pluripotent mass of undifferentiated cells and then is differentiated to develop shoot or root meristem, it is termed indirect organogenesis (Duclercq et al. 2011; Ikeuchi et al. 2016). Organogenesis is a species- and genotype-dependent characteristic. Thus, not all plants can perform organogenesis under *in vitro* conditions. For instance, *Arabidopsis* and *Nicotiana* can regenerate through organogenesis, whereas some plants, including legume plants, and *Gossypium* are recalcitrant to it (Che et al. 2002; Duclercq et al. 2011; Liu et al. 2014).

Plant regeneration through organogenesis depends on various external factors, including nutrient availability, phytohormones, and growth regulators (Ikeuchi et al. 2016; Long et al. 2022). Exogenous hormones, including auxin and cytokinin, are essential for organogenesis as they initiate a signaling response, reprogramming the cell molecular pathway. This reprogramming results in the modulation of various genes, transcription factors, and miRNAs, differentiating the plant cells into specific organ meristematic cells and plant morphogenesis (Bernula et al. 2020; Duclercq et al. 2011; Su and Zhang 2014).

Many genes and transcription factors are recorded to have significant roles in organogenesis, including *WOX11*, *WUS*, and *CUC*. Additionally, miRNAs play a crucial role in regulating these development-associated transcription factors and genes, significantly impacting *in vitro* plant regeneration. miRNA transcriptional analysis of six development stages of *Acacia crassicarpa* organogenetic tissue, zygotic embryo, revealed the identification of 189 known and seven novel miRNAs, including miR159, miR160, miR166, and miR167 (Liu et al. 2014). Small RNA transcriptome sequencing during the *Dendrocalamus latiflorus* (Ma bamboo) shoot organogenesis process resulted in the identification of 727 differentially expressed miRNAs, including miR156, miR164, miR319, and miR398 (Wang et al. 2023).

The ratio of auxin and cytokinin phytohormone has a significant role in shoot organogenesis. Plants in response to auxin activates a group of transcription factors called auxin response factors (ARFs) that bind to auxin response elements (AuxREs), in the promoter region of auxin-responsive genes and regulates their expression (Li et al. 2016). ARFs are vital mediators of auxin signaling during shoot organogenesis, and auxin signaling-associated miRNAs also play a crucial role in *in vitro* plant regeneration. Various miRNAs including miR160, miR167, miR390, and miR393 are reported to be involved in the regulation of auxin signaling (Jodder 2020). miR160 expression was downregulated in 10-day-old totipotent calli kept on shoot induction medium compared to non-totipotent calli of *Arabidopsis,* suggesting its crucial role in shoot organogenesis (Qiao et al. 2012). miR160 targets ARF10 transcription factor and miR160-resistant ARF10 mutant displayed increased callus initiation, shoot organogenesis potential, and exhibited increased expression of *CLV, CUP1, CUP2,* and *WUS* genes (Qiao et al. 2012; Liu et al. 2016). Moreover, ARF10 is reported to repress the *ARR15,* a cytokinin signaling gene indicating co-ordination in auxin and cytokinin signaling (Liu et al. 2016) (Figure 5.1).

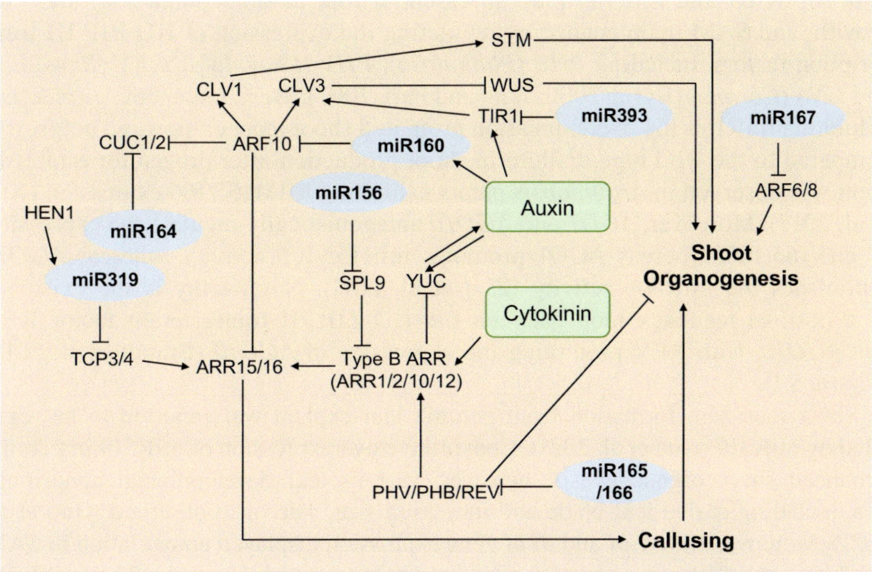

FIGURE 5.1 Representation of miRNAs and gene network that is involved in shoot organogenesis.

miR167 is another auxin signaling-associated miRNA that targets ARF6 and ARF8 transcription factors. Expression of miR167 target mimic in *Nicotiana* resulted in an increased number of shoot initials from leaf explants, suggesting a negative role of miR167 in shoot organogenesis in tobacco (Makkar et al. 2022). Abnormalities in *in vitro* shoot regeneration were observed in *Arabidopsis* plants that exhibited miR393 overexpression as it impacts auxin signaling by targeting *TIR1*, a protein that helps in auxin transport (Wang et al. 2018b) (Figure 5.1).

miR156 significantly contributes to establishing shoot progenitor in shoot promeristem calli (Shin et al. 2019). *Arabidopsis* and tobacco plants exhibiting miR156 overexpression demonstrated promotion in shoot regeneration from leaf explants, whereas reduced expression of miR156 through target mimic resulted in decreased shoot regeneration (Zhang et al. 2015). miR156 indirectly influences the expression of ARR transcription factors, including ARR1, ARR2, and ARR10, which activates cytokinin signaling and *WUS* gene expression, which promotes shoot regeneration through regulation of cell division and is an indicator of shoot progenitor region (Negin et al. 2017; Zhang et al. 2017; Gordon et al. 2007) (Figure 5.1). The expression of miR156 shows age-dependent expression, where with age, miR156 expression decreases and expression of its target gene, SPL9, increases (Wu et al. 2009). The SPL9 transcription factors repress the expression of type-B ARR protein, and thus, an increase in cytokinin signaling is observed in young plants as compared to adult plants (Wu et al. 2009; Zhang et al. 2015; Shin et al. 2019). In Ma bamboo, an increased miR156 expression displayed improved shoot regeneration efficiency from shoot callus, whereas miR156 downregulation resulted in less shoot regeneration (Wang et al. 2023).

The establishment of shoot progenitor is followed by the development of shoot apical meristem (SAM), which is further differentiated to form leaves (Motte et al. 2014). miR165 and miR166 play an essential role in shoot induction, leaf outgrowth, and SAM maintenance by regulating the expression of HD-ZIP III transcription factors, including *PHB (Phabulosa), PHV (Phavoluta), REV (Revoluta),* and *CNA (Corona)* (Byrne 2006; Jung and Park 2007). *Men1* mutant of *Arabidopsis* exhibiting miR165/166 overexpression promoted shoot regeneration and outgrowth compared to the wild type. Failure in shoot production after progenitor establishment was observed in *Arabidopsis* plants exhibiting MIM165/166 expression (Xue et al. 2017). Moreover, *AGO1* and *AGO10* antagonistically regulate the expression of miR165/166, whereas AGO1 promotes miR165/166 activity, whereas *AGO10* diminishes the miRNA activity (Zhu et al. 2011). Noteworthy is the existence of a positive feedback loop between the HD-ZIP III transcription factor REV and *AGO10*, with REV promoting the expression of *AGO10* (Brandt et al. 2013) (Figure 5.1).

Shoot meristem formation from petunia leaf explant was reported to be regulated by miR319 (Guo et al. 2021). Constitutive overexpression of miR319 in petunia enhanced shoot organogenesis; however, various leaf developmental abnormalities, including curled leaf blade and increased size, were also observed (Guo et al. 2021). Moreover, the *NAM* and *STM* gene expression displayed upregulation in SAM exhibiting miR319 overexpression compared to the control (Guo et al. 2021). miR319 also regulates cytokinin signaling by targeting the transcription factors TCP3 and

TCP4, which, in turn, activate the expression of ARR16. The HEN1 protein regulates miR319 expression, and *hen1* mutation resulted in a decrease in miR319 activity and a reduction in shoot regeneration (Yang et al. 2020) (Figure 5.1).

The establishment of SAM and organ boundary determination is regulated by genes belonging to the NAC transcription factor family, including CUC1 and CUC2, which are the targets of miR164 (Nikovics et al. 2006; Wang et al. 2021). Notably, research has indicated the significant involvement of miR171, which targets *HAM* (*Hairy Meristem*), during the development of SAM across various plant species such as *Arabidopsis*, tomato, rice, and barley (Han and Zhou 2022). The overexpression of miR171 in *Arabidopsis,* rice, and tomato generated abnormalities in the growth and development of SAM and caused changes in the expression of meristem regulators such as *CLV1, CLV3*, and *WOX4* (Han and Zhou 2022).

The root organogenesis process *in vitro* plant regeneration from explants leads to the formation of adventitious root (AR). The explants are exposed to certain environmental conditions that induce the cells, similar to pericycle cells of the primary root, to generate root primordia, which differentiate and develop into root apical meristem (RAM). The AR exhibits histological and developmental characteristics with lateral roots, and their induction depends on the genotype of plant and phytohormones, especially auxin and cytokinin, where auxin acts as a promoter and cytokinin as its antagonist (Verstraeten et al. 2014). In the founder cell, during the establishment of root primordia, GATA23, a GATA-type transcription factor, is specifically expressed and is regulated by auxin signaling-associated transcription factors, including IAA28, ARF7, and ARF18 (De Rybel et al. 2010). Moreover, the miR167-ARF6/8 node also influences the activity of GATA23, thus acting as a positive regulator of AR development (Lavenus et al. 2013) (Figure 5.2). Moreover, *arf6* and *arf8 Arabidopsis* double-mutant displayed less AR (Gutierrez et al. 2012). Conversion of root primordia into root meristem depends on auxin accumulation through PIN-mediated transport (Verstraeten et al. 2014). miR393 regulates auxin transport among plant cells *via* targeting TIR1 (Transport Inhibitor Response I) and AFB1/2/3 (Auxin signaling F-box 1/2/3), and overexpression of miR393 in rice resulted in increased lateral root development (Bian et al. 2012; Jiang et al. 2022). The increased expression of *ARF17,* a target of miR160 in the *ago1 Arabidopsis* mutant, decreased AR development (Gutierrez et al. 2012) (Figure 5.2). Moreover, ARF6, ARF8, and ARF17 regulate the expression of the *GH3 (Gretchen Hagel 3*) gene, which is involved in jasmonate signaling and converts active jasmonic acid (JA) into JA-lle. The active JA is an inhibitor of AR, whereas JA-lle inhibits the signaling pathway that represses AR (Gutierrez et al. 2012). Thus, it can be concluded that miR160 and miR167 are essential components of the auxin-JAs crosstalk pathway during AR development. miR165/miR166 also regulates root organogenesis in *Arabidopsis* by regulating RAM patterning, cell differentiation, and maintaining cytokinin homeostasis (Bertolotti et al. 2021). Moreover, it was observed that miR165/166 was activated by SHR (short hair) and SCR (scarecrow) protein (Carlsbecker et al. 2010) (Figure 5.2).

miR390 is also reported to contribute significantly in AR and lateral root development in *Arabidopsis* and *Medicago truncatula* by governing the biogenesis of ta-siRNA, TAS3, which further regulates expression of ARF2/3/4 transcription factor

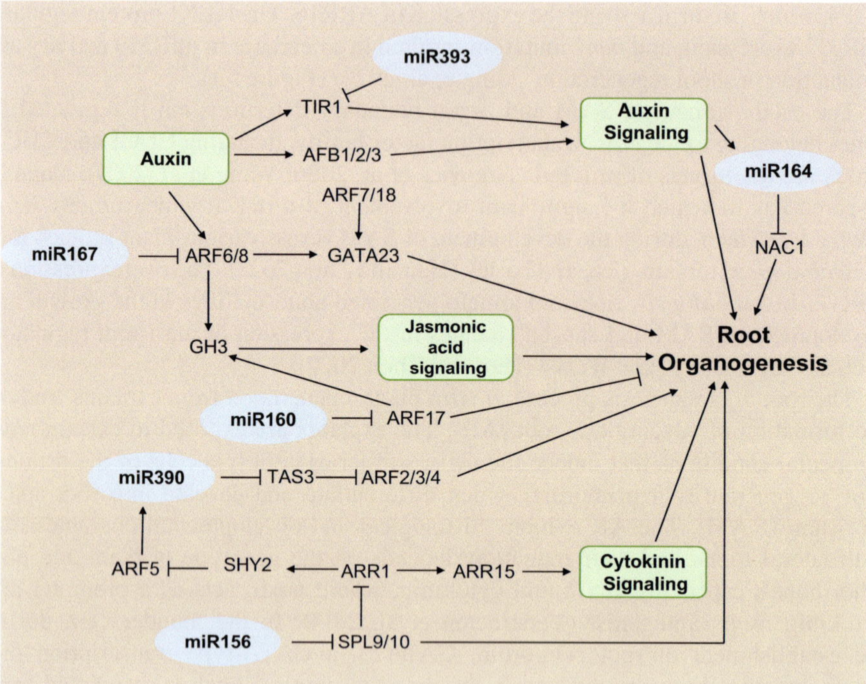

FIGURE 5.2 Representation of miRNAs and gene network that is involved in root organogenesis.

and auxin gradient (Hobecker et al. 2017; Marin et al. 2010). miR156-SPL10 module significantly contributes to root organogenesis by regulating the establishment of root meristem and primary root growth (Barrera-Rojas et al. 2020) (Figure 5.2). *spl10 Arabidopsis* mutant exhibited decreased root meristem activity, whereas the same increased in plants expressing miR156-resistant *SPL10* (Barrera-Rojas et al. 2020). In the root, miR156 expression is influenced by auxin, and similar to shoot meristem, the miR156-SPL10 module is associated with cytokinin signaling as SPL10 represses the expression of the *ARR1* gene (Barrera-Rojas et al. 2020; Yu et al. 2015). These observations indicate that miR156 is an integral component of a complex regulatory circuit that allows crosstalk between auxin and cytokinin hormones (Figure 5.2).

In *Arabidopsis,* root meristem size is regulated by miR396 and its target growth regulating factors (GRFs). Moreover, GRFs repress the *PLT (Plethora)* gene, which regulates cell-cycle-related genes, including *CYCB1 and CYCB2*, and determine stem cell identity. Therefore, miR396 overexpression resulted in the enlarged root meristem, whereas its suppression or target gene overexpression resulted in smaller root meristem size in *Arabidopsis* (Rodriguez et al. 2010, 2015; Rodriguez et al. 2010). In grapes, miR171 is reported to play a significant role during AR formation in *in vitro* grown plantlets, as plants expressing miR171d displayed increased AR formation (Chen et al. 2020).

5.3 miRNA REGULATION OF PLANT SOMATIC EMBRYOGENESIS

SE is another method for *in vitro* plant regeneration where the somatic cell, in response to external stimuli, acquires totipotency and dedifferentiates into embryonic stem cells primarily through embryogenic callus formation. It is a multistep coordinated process that initiates the reprogramming of cell gene expression through the activation of various signaling pathways in response to external stress and phytohormone exposure, including auxin and cytokinin (Marilyn and Harada 1993; Fehér et al. 2003; Dudits et al. 1975). The somatic cell first generates embryogenic callus/pro-embryogenic masses (EC), followed by the differentiation of callus cell into the somatic embryos. Similar to zygotic embryogenesis, distinct development stages are observed during somatic embryo development, including globular, heart, and torpedo (Filonova et al. 2000). Interestingly, it was observed that monocot and dicot plants exhibit different stages of embryo development, where formation of globular, scutella, and coleoptilar stages are observed in monocot plants; and globular, heart-shaped, and torpedo-shaped stages in dicot plant SE (Chaudhary et al. 2003; Yang and Zhang 2010).

The SE is a complex and tightly regulated process where the conversion of somatic cells into embryogenic cells requires differential expression of various genes, transcription factors, and small RNAs associated with various cellular and molecular pathways, including cell division and differentiation, hormone signaling, and cell wall development (Dudits et al. 1975; Fehér et al. 2003). Moreover, this complicated process also depends on various internal and external factors, including the plant's genotype, age, type of explants, nutrients, and phytohormones provided in a culture medium (Siddiqui et al. 2019; Yang et al. 2013a).

miRNA regulates SE by regulating transcription factors and genes involved in molecular switches that convert somatic cells to embryogenic cells. Various miRNAs have been reported to play a significant role in attaining embryogenic potential, developing somatic embryos, and germinating into plantlets. Analysis of 190 miRNAs during SE induction (early and late induction) in *Arabidopsis* revealed differentiation expression of 122 miRNAs, including miR156, miR159, miR398, miR396, miR390, miR157, miR319, and miR167 (Szyrajew et al. 2017). miRNA expression analyses by deep sequencing in three developmental stages of maize embryo-derived embryogenic calli revealed differential expression of several miRNAs involved in the phytohormone signaling pathway. Most of these miRNAs were upregulated in dedifferentiating callus, including miR528, miR156, miR408, miR390, and miR168, whereas miR166 and miR167 were downregulated (Shen et al. 2012).

miRNA expression profiling during SE in *Dimocarpus longan* through deep sequencing revealed more than 600 miRNAs involved. During early embryogenic callus development, miR166 and miR156 family members displayed prominent expression. Similarly, expression of miR26, miR160, miR159, miR390, and miR398 was prominent in heart-shaped and torpedo-stage somatic embryos, whereas in the development of cotyledon embryos, miR4a, miR24, miR167, miR168, miR397, miR398, miR808, and miR5077 contribute significantly (Lin and Lai 2013).

miRNA expression analysis during SE in *Coffea canephora* identified various miRNAs involved in somatic embryo development stages. miR164, miR535, miR168, and miR2119 were reported to be involved in globular somatic embryo formation, while during heart-shaped embryo formation, expression of miR397, miR152, miR2119, miR393, and miR398 was documented. Moreover, it was also revealed that miR157, miR164, miR535, and miR2119 have a significant role in generating totipotent plant cells (Hernández-Castellano et al. 2022). miRNAs exhibiting differential expression during the cotton SE process were identified through high-throughput small RNA and degradome sequencing. Several miRNAs including miR390, miR394, miR397, miR399, miR408, miR829, and miR827 were upregulated in embryogenic callus, whereas miR156, miR2948, miR2949, miR3476, miR162, miR166, miR2950, miR479, miR482, and miR172 displayed down-expression (Yang et al. 2013b).

Phytohormone signaling and its crosstalk with associated signaling pathways are significant in the SE developmental process. Several studies have revealed that miRNAs significantly regulate hormone signaling and phytohormone crosstalk (Jin et al. 2020). Among all phytohormones, auxin is reported to be the foremost hormone that is required for SE induction by activating the auxin signaling pathway that encourages cellular, molecular, and morphological adjustments in the cells (Fehér et al. 2003; Yang and Zhang 2010). Furthermore, cytokinin, gibberellic acid (GA), ethylene, abscisic acid (ABA), and JAs hormones also contribute significantly to the induction and progression of SE. Various miRNAs are identified as regulators of this hormone-signaling pathway and are also involved in phytohormone crosstalk (Jin et al. 2020). miR160, miR166, miR167, miR393, and miR390 regulate auxin signaling and its crosstalk with other hormone signaling.

miR160 negatively regulates auxin signaling by suppressing the expression of ARF10, ARF17, and ARF16 and regulates plant organ development, including root, shoot, and embryo, and is also involved in seed germination (Lin et al. 2015a). In *D. longan*, miR160 demonstrated maximum expression in the developmental stages of somatic embryos (torpedo, globular, and cotyledon embryos) compared to embryogenic callus (Lin et al. 2015a). Moreover, it was recorded that in response to salicylic acid, miR160 exhibited down-expression, whereas it displayed upregulation in response to GA, ethylene, and JA. These results suggest that miR160 is involved in auxin, GA, ethylene, and JA crosstalk during *D. longan* SE and somatic embryo development (Lin et al. 2015a). During SE of *Picea abies* and *Pinus pinaster*, miR160 expression increased gradually from embryogenic callus to mature somatic embryos, whereas its target, *ARF18,* demonstrated gradual down-expression (Alves et al. 2022; Rodrigues et al. 2019). Moreover, miR160 is also recognized as a moderator of auxin and cytokinin signaling, as ARF10 represses ARR15, a regulator of cytokinin signaling (Jin et al. 2020) (Figure 5.3).

Interestingly, the miR160-ARF10/16/17 module regulates the expression of *LEC2,* a crucial gene required for SE induction, and activates *YUC* genes involved in auxin biosynthesis and transport (Wójcikowska et al. 2013). The *LEC2* gene expression was upregulated in *Arabidopsis* plants with decreased miR160 expression or exhibiting miR160-resistant ARF16 expression. Also, the expression of

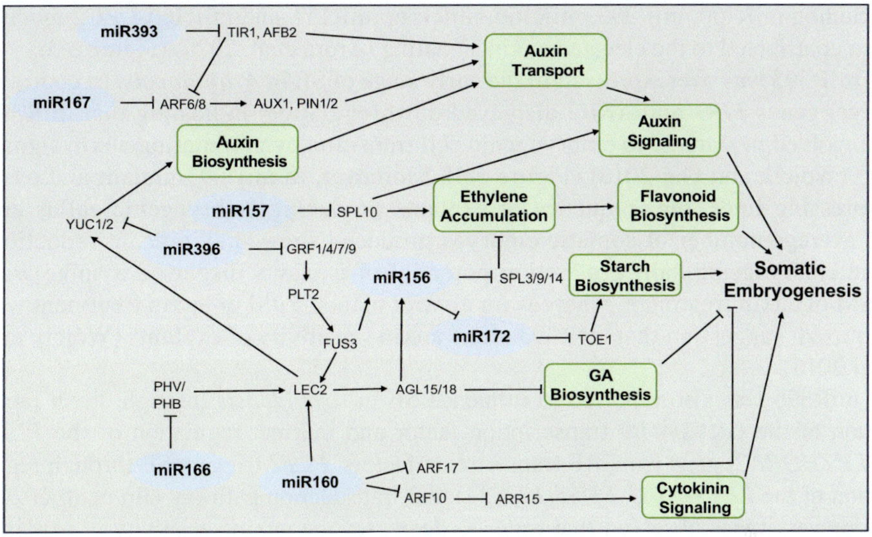

FIGURE 5.3 miRNAs regulate somatic embryogenesis through modulating hormone signaling, and flavonoid and starch biosynthesis.

YUC2/4/10, *IAA1/17*, and *ACS* genes were upregulated (Figure 5.3). Similarly, miR160 upregulation and *arf10/arf16* mutants displayed down-expression of *LEC2* (Wójcik et al. 2017).

Moreover, *LEC2* expression is also influenced by PHB/PHV transcription factors, which are targets of miR166. Thus, miR160 and miR166 influence SE induction and embryo development through modulating auxin signaling (Wójcik et al. 2017). Effect of miR166 on auxin signaling was also observed in *Larix leptolepsis,* where its overexpression led to upregulation of *NIT (Nitrilase), ARF1/2,* and exhibited high somatic embryo germination rate (Li et al. 2017, 2018). The expression of miR166 displayed gradual down-expression from the early embryogenic callus to the globular stage embryo of *D. longan*, whereas its target gene *ATHB15* displayed upregulation (Zhang et al. 2020).

In *Arabidopsis,* overexpression of miR167 inhibited SE through negative regulation of auxin signaling in embryogenic callus (Su et al. 2016). In cotton SE, miR167 expression displayed significant expression in non-embryogenic callus (NEC) and at initiation of callusing. As the SE process progressed, a significant reduction in miR167 expression was observed until the torpedo-stage embryo; a sudden spike in expression was recorded in the cotyledon-stage embryo (Yang et al. 2013b). Moreover, during cotton SE, expression of target mimic sequence of miRNA167 (MIM167) enhanced cellular auxin signaling indicated by increased expression of *ARF6/8, Aux1, LAX3, and PIN1/2* genes, which resulted in increased callogenesis and somatic embryo formation in cotton (Arora et al. 2020). Moreover, it was revealed that the diminution of miR167 led to misexpression of other associated miRNAs, including miR160, miR393, miR166, miR156, miR157, and miR166-LEC2 module also contributed to the elevated auxin signaling (Arora et al. 2020) (Figure 5.3).

miR393 was overexpressed in the early stage of SE in *Arabidopsis*. In contrast, target genes *TIR1* and *AFB2* displayed downregulation, indicating that miR393 is involved in somatic to embryogenic cell transition by modulating auxin signaling (Wójcik and Gaj 2016) (Figure 5.3). Moreover, in miR393 mutant and overexpressing lines, the frequency of explants producing embryogenic callus and an average number of somatic embryos produced showed significant reduction. The embryogenic calli production potential of explants displayed a spike with 1 μM of auxin treatment, whereas for control plants, 5 μM of auxin treatment was required, suggesting that miR393 alters auxin sensitivity of explants (Wójcik and Gaj 2016).

miR396 was also reported to influence SE in *Arabidopsis* through direct regulation of the GRF1/4/7/9 transcription factor and indirect regulation of the *PLT2 (PLETHORA2)* gene *via* GRF transcription factors. *PLT2* triggers SE through regulation of the *LEC1-ABI3-FUS3-LEC2* signal transduction pathway (Jin et al. 2020). Moreover, it was observed that miR396 decreases the auxin sensitivity of explants (Szczygieł-Sommer and Gaj 2019). In *miR396* mutants, an increase in IAA level and *YUC1/2* expression was recorded, whereas the same displayed downregulation in miR396 overexpressing lines, suggesting miR396 also affects auxin signaling during SE (Figure 5.3).

As discussed, miR390 regulates auxin signaling through cleavage of ARF3/4 *via* ta-siRNA and is involved in root formation. Interestingly, miR390 has a significant role during SE, as evidenced by various miRNA profiling studies (Lin and Lai 2013; Szyrajew et al. 2017). In *Arabidopsis* and cotton embryogenic callus, miR390 expression was upregulated in the early stage of embryogenic callus, followed by a reduction in expression in later stages of embryo development (Szyrajew et al. 2017; Wang et al. 2023). However, in *D. longan*, miR160 expression displayed different expression patterns, with low expression in embryogenic callus and highest expression in torpedo-stage somatic embryos (Lin et al. 2015b)

In *Fortunella hindsii* (wild kumquat), overexpression of miR156 and knockout of its target genes, *SPL3* and *SPL14,* enhanced somatic embryo production (Long et al. 2018). miR156 expression is regulated by AGL15 and FUS3 transcription factors. Moreover, in embryogenic callus, expression of MIM156 resulted in down-expression of miR172 and upregulation of the *TOE1* gene, which led to reduced expression of genes involved in starch biosynthesis and transport. Thus, in the callus with MIM156 expression, the amyloplast and starch content was significantly lower than the control callus, negatively affecting the SE ability (Feng et al. 2022) (Figure 5.3). During cotton SE, miR156 expression displayed upregulation throughout the process, where the expression was least in embryogenic callus and maximum in cotyledon embryos (Yang et al. 2013b). In *Arabidopsis*, miR156 was upregulated in the embryogenic callus, whereas, in coconut, miR156 was downregulated in the embryogenic callus, suggesting that miR156 expression varies among monocot and dicot plants during SE (Jin et al. 2020; Szyrajew et al. 2017). Notably, the miR156-SPL9 module regulates SE in Japanese Larch by responding to ABA and is involved in crosstalk between GA and SA (Zhang et al. 2021).

miR157 expression was significantly upregulated in NEC and during the early stage of callus initiation. In contrast, in EC and during somatic embryo development, miR157 exhibited down-expression in cotton (Wang et al. 2018a). miR157 regulates SPL10 in cotton and overexpression of *SPL10* resulted in accumulation of free auxin and ethylene hormones, activation of their signaling pathway, flavonoid biosynthesis, and displayed increased callus proliferation and rate of dedifferentiation (Wang et al. 2018a) (Figure 5.3). miR171 regulates SE in citrus plants by repressing SCL (Scarecrow-like) transcription factors that activate genes and processes (biological and metabolic) associated with cell differentiation. miR171 overexpression in recalcitrant callus resulted in SE acquisition, whereas its reduction through target mimics decreased SE competence in strong embryogenic callus (Shi et al. 2022).

5.4 CONCLUSION

In conclusion, small RNAs have emerged as critical regulators of plant regeneration, influencing various aspects of plant developmental biology, including cell proliferation, differentiation, and tissue formation. Small RNAs, such as miRNAs and siRNAs, are crucial in orchestrating gene expression, regulating developmental events, and enabling plants to adapt to stress responses during regeneration. Understanding the complex network of small-RNA-mediated regulation in plant regeneration provides insights into fundamental aspects of plant biology and holds promising implications for improving agricultural and biotechnological contexts. Manipulation of small RNA pathways emerges as a potential strategy to enhance plant propagation and crop productivity. Further research in this field will help develop efficient and sustainable plant regeneration techniques to meet the growing demand for food and other plant-derived products.

REFERENCES

Alves, Ana, Ana Confraria, Susana Lopes, Bruno Costa, Pedro Perdiguero, Ana Milhinhos, Elena Baena-González, Sandra Correia, and Célia M. Miguel. 2022. "miR160 interacts *in vivo* with *Pinus pinaster AUXIN RESPONSE FACTOR 18* target site and negatively regulates its expression during conifer somatic embryo development." *Front Plant Sci* 13. https://doi.org/10.3389/fpls.2022.857611.

Arora, Sakshi, Amarjeet Kumar Singh, and Bhupendra Chaudhary. 2020. "Target-mimicry based miRNA167-diminution ameliorates cotton somatic embryogenesis via transcriptional biases of auxin signaling associated miRNAs and genes." *Plant Cell Tissue Organ Cult* 141 (3): 511–31. https://doi.org/10.1007/s11240-020-01810-9.

Barrera-Rojas, Carlos Hernán, Gabriel Henrique Braga Rocha, Laura Polverari, Diego Armando Pinheiro Brito, Diego Silva Batista, Marcela M. Notini, Ana Claudia Ferreira da Cruz, Edna Gicela Ortiz Morea, Sabrina Sabatini, Wagner Campos Otoni, and Fabio Tebaldi Silveira Nogueira. 2020. "miR156-targeted *SPL10* controls Arabidopsis root meristem activity and root-derived *de novo* shoot regeneration via cytokinin responses." *J Exp Bot* 71 (3): 934–50. https://doi.org/10.1093/jxb/erz475.

Bartel, David P. 2004. "MicroRNAs: genomics, biogenesis, mechanism, and function." *Cell* 116 (2): 281–97. https://doi.org/10.1016/S0092-8674(04)00045-5.

Baumberger, N., and D. C. Baulcombe. 2005. "*Arabidopsis* ARGONAUTE1 is an RNA Slicer that selectively recruits microRNAs and short interfering RNAs." *Proc Natl Acad Sci USA* 102 (33): 11928–33. https://doi.org/10.1073/pnas.0505461102.

Bernula, Dóra, Péter Benkő, Nikolett Kaszler, Ildikó Domonkos, Ildikó Valkai, Réka Szőllősi, Györgyi Ferenc, Ferhan Ayaydin, Attila Fehér, and Katalin Gémes. 2020. "Timely removal of exogenous cytokinin and the prevention of auxin transport from the shoot to the root affect the regeneration potential of Arabidopsis roots." *Plant Cell Tissue Organ Cult* 140 (2): 327–39. https://doi.org/10.1007/s11240-019-01730-3.

Bertolotti, Gaia, Daria Scintu, and Raffaele Dello Ioio. 2021. "A small cog in a large wheel: crucial role of miRNAs in root apical meristem patterning." *J Exp Bot* 72 (19): 6755–67. https://doi.org/10.1093/jxb/erab332.

Bian, Hongwu, Yakun Xie, Fu Guo, Ning Han, Shengyun Ma, Zhanghui Zeng, Junhui Wang, Yinong Yang, and Muyuan Zhu. 2012. "Distinctive expression patterns and roles of the miRNA393/TIR1 homolog module in regulating flag leaf inclination and primary and crown root growth in rice (*Oryza sativa*)." *New Phytol* 196 (1): 149–61. https://doi.org/10.1111/j.1469-8137.2012.04248.x.

Bidabadi, Siamak Shirani, and S. Mohan Jain. 2020. "Cellular, molecular, and physiological aspects of in vitro plant regeneration." *Plants* 9 (6): 702. https://www.mdpi.com/2223-7747/9/6/702.

Brandt, Ronny, Yakun Xie, Thomas Musielak, Moritz Graeff, York-Dieter Stierhof, Hai Huang, Chun-Ming Liu, and Stephan Wenkel. 2013. "Control of stem cell homeostasis via interlocking microRNA and microProtein feedback loops." *Mech Dev* 130 (1): 25–33. https://doi.org/10.1016/j.mod.2012.06.007.

Byrne, Mary E. 2006. "Shoot meristem function and leaf polarity: the role of class III HD–ZIP genes." *PLOS Genet* 2 (6): e89. https://doi.org/10.1371/journal.pgen.0020089.

Carlsbecker, Annelie, Ji-Young Lee, Christina J. Roberts, Jan Dettmer, Satu Lehesranta, Jing Zhou, Ove Lindgren, Miguel A. Moreno-Risueno, Anne Vatén, Siripong Thitamadee, Ana Campilho, Jose Sebastian, John L. Bowman, Ykä Helariutta, and Philip N. Benfey. 2010. "Cell signalling by microRNA165/6 directs gene dose-dependent root cell fate." *Nature* 465 (7296): 316–21. https://doi.org/10.1038/nature08977.

Chaudhary, B., S. Kumar, K. V. Prasad, G. S. Oinam, P. K. Burma, and D. Pental. 2003. "Slow desiccation leads to high-frequency shoot recovery from transformed somatic embryos of cotton (*Gossypium hirsutum* L. cv. Coker 310 FR)." *Plant Cell Rep* 21 (10): 955–60. https://doi.org/10.1007/s00299-003-0613-x.

Che, Ping, Derek J. Gingerich, Sonia Lall, and Stephen H. Howell. 2002. "Global and hormone-induced gene expression changes during shoot development in Arabidopsis." *Plant Cell* 14 (11): 2771–85. https://doi.org/10.1105/tpc.006668.

Chen, Qiu-ju, Bo-han Deng, Jie Gao, Zhong-yang Zhao, Zi-li Chen, Shi-ren Song, Lei Wang, Li-ping Zhao, Wen-ping Xu, Cai-xi Zhang, Chao Ma, and Shi-ping Wang. 2020. "A miRNA-encoded small peptide, vvi-miPEP171d1, regulates adventitious root formation." *Plant Physiol* 183 (2): 656–70. https://doi.org/10.1104/pp.20.00197.

Chen, Xuemei. 2005. "MicroRNA biogenesis and function in plants." *FEBS Lett* 579 (26): 5923–31. https://doi.org/10.1016/j.febslet.2005.07.071.

De Rybel, Bert, Valya Vassileva, Boris Parizot, Marlies Demeulenaere, Wim Grunewald, Dominique Audenaert, Jelle Van Campenhout, Paul Overvoorde, Leentje Jansen, Steffen Vanneste, Barbara Möller, Michael Wilson, Tara Holman, Gert Van Isterdael, Géraldine Brunoud, Marnik Vuylsteke, Teva Vernoux, Lieven De Veylder, Dirk Inzé, Dolf Weijers, Malcolm J. Bennett, and Tom Beeckman. 2010. "A novel Aux/IAA28 signaling cascade activates GATA23-dependent specification of lateral root founder cell identity." *Curr Biol* 20 (19): 1697–706. https://doi.org/10.1016/j.cub.2010.09.007.

Duclercq, Jérôme, Brigitte Sangwan-Norreel, Manuella Catterou, and Rajbir S. Sangwan. 2011. "*De novo* shoot organogenesis: from art to science." *Trends Plant Sci* 16 (11): 597–606. https://doi.org/10.1016/j.tplants.2011.08.004.

Dudits, Dénes, László Bögre, and Györgyey János. 1975. "Molecular and cellular approaches to the analysis of plant embryo development from somatic cells in vitro." *J Cell Sci* 99 (3): 473–82. https://doi.org/10.1242/jcs.99.3.473.

Fehér, Attila, Taras P. Pasternak, and Dénes Dudits. 2003. "Transition of somatic plant cells to an embryogenic state." *Plant Cell Tissue Organ Cult* 74 (3): 201–28. https://doi.org/10.1023/A:1024033216561.

Feng, Meng-Qi, Meng-Di Lu, Jian-Mei Long, Zhao-Ping Yin, Nan Jiang, Peng-Bo Wang, Yue Liu, Wen-Wu Guo, and Xiao-Meng Wu. 2022. "miR156 regulates somatic embryogenesis by modulating starch accumulation in citrus." *J Exp Bot* 73 (18): 6170–85. https://doi.org/10.1093/jxb/erac248.

Filonova, Lada H., Peter V. Bozhkov, Vladimir B. Brukhin, Geoffrey Daniel, Boris Zhivotovsky, and Sara von Arnold. 2000. "Two waves of programmed cell death occur during formation and development of somatic embryos in the gymnosperm, Norway spruce." *J Cell Sci* 113 (Pt 24): 4399–411. https://doi.org/10.1242/jcs.113.24.4399.

Gordon, Sean P., Marcus G. Heisler, G. Venugopala Reddy, Carolyn Ohno, Pradeep Das, and Elliot M. Meyerowitz. 2007. "Pattern formation during de novo assembly of the Arabidopsis shoot meristem." *Development* 134 (19): 3539–48. https://doi.org/10.1242/dev.010298.

Guo, Yulong, Xiaoting Qin, Bin Zhang, Xiaojing Xu, Zhineng Li, and Mingyang Li. 2021. "Overexpression of miR319 in petunia (*Petunia×hybrida*) promotes *de novo* shoot organogenesis from leaf explants." *In Vitro Cell Dev Biol Plant* 57 (1): 72–9. https://doi.org/10.1007/s11627-020-10063-2.

Gutierrez, Laurent, Gaëlle Mongelard, Kristýna Floková, Daniel I. Păcurar, Ondřej Novák, Paul Staswick, Mariusz Kowalczyk, Monica Păcurar, Hervé Demailly, Gaia Geiss, and Catherine Bellini. 2012. "Auxin controls arabidopsis adventitious root initiation by regulating jasmonic acid homeostasis." *Plant Cell* 24 (6): 2515–27. https://doi.org/10.1105/tpc.112.099119.

Han, Han, and Yun Zhou. 2022. "Function and regulation of microRNA171 in plant stem cell homeostasis and developmental programing." *Int J Mol Sci* 23 (5). https://doi.org/10.3390/ijms23052544.

Hernández-Castellano, Sara, Mauricio Andrade-Marcial, Edder D. Aguilar-Méndez, Víctor M. Loyola-Vargas, Stefan de Folter, and Clelia De-la-Peña. 2022. "MiRNA expression analysis during somatic embryogenesis in *Coffea canephora*." *Plant Cell Tissue Organ Cult* 150 (1): 177–90. https://doi.org/10.1007/s11240-022-02258-9.

Hill, Kristine, and G. Eric Schaller. 2013. "Enhancing plant regeneration in tissue culture: a molecular approach through manipulation of cytokinin sensitivity." *Plant Signal Behav* 8 (10). https://doi.org/10.4161/psb.25709.

Hobecker, Karen Vanesa, Mauricio Alberto Reynoso, Pilar Bustos-Sanmamed, Jiangqi Wen, Kirankumar S. Mysore, Martín Crespi, Flavio Antonio Blanco, and María Eugenia Zanetti. 2017. "The MicroRNA390/TAS3 pathway mediates symbiotic nodulation and lateral root growth." *Plant Physiol* 174 (4): 2469–86. https://doi.org/10.1104/pp.17.00464.

Ikeuchi, Momoko, Yoichi Ogawa, Akira Iwase, and Keiko Sugimoto. 2016. "Plant regeneration: cellular origins and molecular mechanisms." *Development* 143 (9): 1442–51. https://doi.org/10.1242/dev.134668.

Jiang, Jinjin, Haotian Zhu, Na Li, Jacqueline Batley, and Youping Wang. 2022. "The miR393-target module regulates plant development and responses to biotic and abiotic stresses." *Int J Mol Sci* 23 (16). https://doi.org/10.3390/ijms23169477.

Jin, Longfei, Rajesh Yarra, Lixia Zhou, Zhihao Zhao, and Hongxing Cao. 2020. "miRNAs as key regulators via targeting the phytohormone signaling pathways during somatic embryogenesis of plants." *3 Biotech* 10 (11): 495. https://doi.org/10.1007/s13205-020-02487-9.

Jodder, Jayanti. 2020. "miRNA-mediated regulation of auxin signaling pathway during plant development and stress responses." *J Biosci* 45 (1): 91. https://doi.org/10.1007/s12038-020-00062-1.

Jung, Jae-Hoon, and Chung-Mo Park. 2007. "MIR166/165 genes exhibit dynamic expression patterns in regulating shoot apical meristem and floral development in Arabidopsis." *Planta* 225 (6): 1327–38. https://doi.org/10.1007/s00425-006-0439-1.

Jung, Jae-Hoon, Pil Joon Seo, and Chung-Mo Park. 2009. "microRNA biogenesis and function in higher plants." *Plant Biotechnol Rep* 3 (2): 111–26. https://doi.org/10.1007/s11816-009-0085-8.

Lardon, Robin, and Danny Geelen. 2020. "Natural variation in plant pluripotency and regeneration." *Plants*, 9 (10): 1261. https://www.mdpi.com/2223-7747/9/10/1261.

Lavenus, Julien, Tatsuaki Goh, Ianto Roberts, Soazig Guyomarc'h, Mikaël Lucas, Ive De Smet, Hidehiro Fukaki, Tom Beeckman, Malcolm Bennett, and Laurent Laplaze. 2013. "Lateral root development in *Arabidopsis* fifty shades of auxin." *Trends Plant Sci* 18 (8): 450–58. https://doi.org/10.1016/j.tplants.2013.04.006.

Li, Si-Bei, Zong-Zhou Xie, Chun-Gen Hu, and Jin-Zhi Zhang. 2016. "A review of auxin response factors (ARFs) in plants." *Front Plant Sci* 7. https://doi.org/10.3389/fpls.2016.00047.

Li, Zhe-Xin, Yan-Ru Fan, Shao-Fei Dang, Wan-Feng Li, Li-Wang Qi, and Su-Ying Han. 2018. "LaMIR166a-mediated auxin biosynthesis and signalling affect somatic embryogenesis in Larix leptolepis." *Mol Genet Genomic Med* 293 (6): 1355–63. https://doi.org/10.1007/s00438-018-1465-y.

Li, Zhe-Xin, Li-Feng Zhang, Wan-Feng Li, Li-Wang Qi, and Su-Ying Han. 2017. "MIR166a affects the germination of somatic embryos in larixleptolepis by modulating IAA biosynthesis and signaling genes." *J Plant Growth Regul* 36 (4): 889–96. https://doi.org/10.1007/s00344-017-9693-7.

Lin, Yuling, and Zhongxiong Lai. 2013. "Comparative analysis reveals dynamic changes in miRNAs and their targets and expression during somatic embryogenesis in longan (*Dimocarpus longan* Lour.)." *PLoS One* 8 (4): e60337. https://doi.org/10.1371/journal.pone.0060337.

Lin, Yuling, Zhongxiong Lai, Qilin Tian, Lixia Lin, Ruilian Lai, Manman Yang, Dongmin Zhang, Yukun Chen, and Zihao Zhang. 2015a. "Endogenous target mimics down-regulate miR160 mediation of *ARF10*, *-16*, and *-17* cleavage during somatic embryogenesis in *Dimocarpus longan* Lour." *Front Plant Sci* 6. https://doi.org/10.3389/fpls.2015.00956.

Lin, Yuling, Lixia Lin, Ruilian Lai, Weihua Liu, Yukun Chen, Zihao Zhang, Xu XuHan, and Zhongxiong Lai. 2015b. "MicroRNA390-directed TAS3 cleavage leads to the production of tasiRNA-*ARF3/4* during somatic embryogenesis in *Dimocarpus longan* Lour." *Front Plant Sci* 6. https://doi.org/10.3389/fpls.2015.01119.

Liu, Weina, Wangning Yu, Lingyu Hou, Xiaoyu Wang, Fei Zheng, Weixuan Wang, Di Liang, Hailun Yang, Yi Jin, and Xiangming Xie. 2014. "Analysis of miRNAs and their targets during adventitious shoot organogenesis of *Acacia crassicarpa*." *PLoS One* 9 (4): e93438. https://doi.org/10.1371/journal.pone.0093438.

Liu, Zhenhua, Juan Li, Long Wang, Qiang Li, Qing Lu, Yanchong Yu, Shuo Li, Ming-yi Bai, Yuxin Hu, and Fengning Xiang. 2016. "Repression of callus initiation by the miRNA-directed interaction of auxin-cytokinin in *Arabidopsis thaliana*." *Plant J* 87 (4): 391–402. https://doi.org/10.1111/tpj.13211.

Long, Jian-Mei, Chao-Yang Liu, Meng-Qi Feng, Yun Liu, Xiao-Meng Wu, and Wen-Wu Guo. 2018. "miR156-SPL modules regulate induction of somatic embryogenesis in citrus callus." *J Exp Bot* 69 (12): 2979–93. https://doi.org/10.1093/jxb/ery132.

Long, Yun, Yun Yang, Guangtang Pan, and Yaou Shen. 2022. "New insights into tissue culture plant-regeneration mechanisms." *Front Plant Sci* 13: 926752. https://doi.org/10.3389/fpls.2022.926752.

Makkar, Harshita, Sakshi Arora, Aniruddhabhai K. Khuman, and Bhupendra Chaudhary. 2022. "Target-mimicry-based miR167 diminution confers salt-stress tolerance during in vitro organogenesis of tobacco (*Nicotiana tabacum* L.)." *J Plant Growth Regul* 41 (4): 1462–80. https://doi.org/10.1007/s00344-021-10376-5.

Marilyn, A. L. West, and John J. Harada. 1993. "Embryogenesis in higher plants: an overview." *The Plant Cell* 5 (10): 1361–9. https://doi.org/10.2307/3869788.

Marin, Elena, Virginie Jouannet, Aurélie Herz, Annemarie S. Lokerse, Dolf Weijers, Herve Vaucheret, Laurent Nussaume, Martin D. Crespi, and Alexis Maizel. 2010. "miR390, Arabidopsis TAS3 tasiRNAs, and their AUXIN RESPONSE FACTOR targets define an autoregulatory network quantitatively regulating lateral root growth." *Plant Cell* 22 (4): 1104–17. https://doi.org/10.1105/tpc.109.072553.

Motte, Hans, Danny Vereecke, Danny Geelen, and Stefaan Werbrouck. 2014. "The molecular path to in vitro shoot regeneration." *Biotechnol Adv* 32 (1): 107–21. https://doi.org/10.1016/j.biotechadv.2013.12.002.

Negin, Boaz, Or Shemer, Yonatan Sorek, and Leor Eshed Williams. 2017. "Shoot stem cell specification in roots by the WUSCHEL transcription factor." *PLoS One* 12 (4): e0176093. https://doi.org/10.1371/journal.pone.0176093.

Nikovics, Krisztina, Thomas Blein, Alexis Peaucelle, Tetsuya Ishida, Halima Morin, Mitsuhiro Aida, and Patrick Laufs. 2006. "The balance between the *MIR164A* and *CUC2* genes controls leaf margin serration in *Arabidopsis*." *Plant Cell* 18 (11): 2929–45. https://doi.org/10.1105/tpc.106.045617.

Qiao, Meng, Zhongjuan Zhao, Yuguang Song, Zhenhua Liu, Lingxue Cao, Yanchong Yu, Shuo Li, and Fengning Xiang. 2012. "Proper regeneration from in vitro cultured Arabidopsis thaliana requires the microRNA-directed action of an auxin response factor." *Plant J* 71 (1): 14–22. https://doi.org/10.1111/j.1365-313X.2012.04944.x.

Rodrigues, Andreia S., Inês Chaves, Bruno Vasques Costa, Yao-Cheng Lin, Susana Lopes, Ana Milhinhos, Yves Van de Peer, and Célia M. Miguel. 2019. "Small RNA profiling in *Pinus pinaster* reveals the transcriptome of developing seeds and highlights differences between zygotic and somatic embryos." *Sci Rep* 9 (1): 11327. https://doi.org/10.1038/s41598-019-47789-y.

Rodriguez, Ramiro E., María Florencia Ercoli, Juan Manuel Debernardi, Natalie W. Breakfield, Martin A. Mecchia, Martin Sabatini, Toon Cools, Lieven De Veylder, Philip N. Benfey, and Javier F. Palatnik. 2015. "MicroRNA miR396 regulates the switch between stem cells and transit-amplifying cells in Arabidopsis roots." *Plant Cell* 27 (12): 3354–66. https://doi.org/10.1105/tpc.15.00452.

Rodriguez, Ramiro E., Martin A. Mecchia, Juan M. Debernardi, Carla Schommer, Detlef Weigel, and Javier F. Palatnik. 2010. "Control of cell proliferation in *Arabidopsis thaliana* by microRNA miR396." *Development* 137 (1): 103–12. https://doi.org/10.1242/dev.043067.

Shen, Yaou, Zhou Jiang, Xiadong Yao, Zhiming Zhang, Haijian Lin, Maojun Zhao, Hailan Liu, Huanwei Peng, Shujun Li, and Guangtang Pan. 2012. "Genome expression profile analysis of the immature maize embryo during dedifferentiation." *PLoS One* 7 (3): e32237. https://doi.org/10.1371/journal.pone.0032237.

Shi, Qiao-Fang, Jian-Mei Long, Zhao-Ping Yin, Nan Jiang, Meng-Qi Feng, Bo Zheng, Wen-Wu Guo, and Xiao-Meng Wu. 2022. "miR171 modulates induction of somatic embryogenesis in citrus callus." *Plant Cell Rep* 41 (6): 1403–15. https://doi.org/10.1007/s00299-022-02865-y.

Shin, Jinwoo, Soonhyung Bae, and Pil Joon Seo. 2019. "De novo shoot organogenesis during plant regeneration." *J Exp Bot* 71 (1): 63–72. https://doi.org/10.1093/jxb/erz395.

Shriram, Varsha, Vinay Kumar, Rachayya M. Devarumath, Tushar S. Khare, and Shabir H. Wani. 2016. "MicroRNAs as potential targets for abiotic stress tolerance in plants." *Front Plant Sci* 7: 817. https://doi.org/10.3389/fpls.2016.00817.

Siddiqui, Zahid Hameed, Zahid Khorshid Abbas, Mohammad Wahid Ansari, and Mohammad Nasir Khan. 2019. "The role of miRNA in somatic embryogenesis." *Genomics* 111 (5): 1026–33. https://doi.org/10.1016/j.ygeno.2018.11.022.

Su, Ying Hua, Yu Bo Liu, Chao Zhou, Xiao Ming Li, and Xian Sheng Zhang. 2016. "The microRNA167 controls somatic embryogenesis in Arabidopsis through regulating its target genes ARF6 and ARF8." *Plant Cell Tissue Organ Cult* 124 (2): 405–17. https://doi.org/10.1007/s11240-015-0903-3.

Su, Ying Hua, and Xian Sheng Zhang. 2014. "The hormonal control of regeneration in plants." In *Current Topics in Developmental Biology*, edited by Brigitte Galliot, pp. 35–69. Academic Press.

Sunkar, Ramanjulu, Yong-Fang Li, and Guru Jagadeeswaran. 2012. "Functions of microRNAs in plant stress responses." *Trends Plant Sci* 17 (4): 196–203. https://doi.org/10.1016/j.tplants.2012.01.010.

Szczygieł-Sommer, Aleksandra, and Małgorzata D. Gaj. 2019. "The miR396-GRF regulatory module controls the embryogenic response in Arabidopsis via an auxin-related pathway." *Int J Mol Sci* 20 (20): 5221. https://www.mdpi.com/1422-0067/20/20/5221.

Szyrajew, Katarzyna, Dawid Bielewicz, Jakub Dolata, Anna M. Wójcik, Katarzyna Nowak, Aleksandra Szczygieł-Sommer, Zofia Szweykowska-Kulinska, Artur Jarmolowski, and Małgorzata D. Gaj. 2017. "MicroRNAs are intensively regulated during induction of somatic embryogenesis in Arabidopsis." *Front Plant Sci* 8. https://doi.org/10.3389/fpls.2017.00018.

Verstraeten, Inge, Sébastien Schotte, and Danny Geelen. 2014. "Hypocotyl adventitious root organogenesis differs from lateral root development." *Front Plant Sci* 5. https://doi.org/10.3389/fpls.2014.00495.

Wang, Jun, Jinlin Bao, Beibei Zhou, Min Li, Xizhi Li, and Jian Jin. 2021. "The *osa-miR164* target *OsCUC1* functions redundantly with *OsCUC3* in controlling rice meristem/organ boundary specification." *New Phytol* 229 (3): 1566–81. https://doi.org/10.1111/nph.16939.

Wang, Lichen, Nian Liu, Tianyi Wang, Jianying Li, Tianwang Wen, Xiyan Yang, Keith Lindsey, and Xianlong Zhang. 2018a. "The GhmiR157a-GhSPL10 regulatory module controls initial cellular dedifferentiation and callus proliferation in cotton by modulating ethylene-mediated flavonoid biosynthesis." *J Exp Bot* 69 (5): 1081–93. https://doi.org/10.1093/jxb/erx475.

Wang, Long, Zhenhua Liu, Meng Qiao, and Fengning Xiang. 2018b. "miR393 inhibits in vitro shoot regeneration in arabidopsis thaliana via repressing TIR1." *Plant Sci* 266: 1–8. https://doi.org/10.1016/j.plantsci.2017.10.009.

Wang, Nannan, Wenjia Wang, Yang Cheng, Changyang Cai, and Qiang Zhu. 2023. "Uncovering the miRNA-mediated regulatory network involved in Ma bamboo (*Dendrocalamus latiflorus*) de novo shoot organogenesis." *Hortic Res* 10 (12). https://doi.org/10.1093/hr/uhad223.

Wójcik, Anna M., and Małgorzata D. Gaj. 2016. "miR393 contributes to the embryogenic transition induced in vitro in Arabidopsis via the modification of the tissue sensitivity to auxin treatment." *Planta* 244 (1): 231–43. https://doi.org/10.1007/s00425-016-2505-7.

Wójcik, Anna M., Michael D. Nodine, and Małgorzata D. Gaj. 2017. "miR160 and miR166/165 contribute to the LEC2-mediated auxin response involved in the somatic embryogenesis induction in Arabidopsis." *Front Plant Sci* 8. https://doi.org/10.3389/fpls.2017.02024.

Wójcikowska, Barbara, Karolina Jaskóła, Przemysław Gąsiorek, Magdalena Meus, Katarzyna Nowak, and Małgorzata D. Gaj. 2013. "LEAFY COTYLEDON2 (LEC2) promotes embryogenic induction in somatic tissues of Arabidopsis, via YUCCA-mediated auxin biosynthesis." *Planta* 238 (3): 425–40. https://doi.org/10.1007/s00425-013-1892-2.

Wu, Gang, Mee Yeon Park, Susan R. Conway, Jia-Wei Wang, Detlef Weigel, and R. Scott Poethig. 2009. "The sequential action of miR156 and miR172 regulates developmental timing in Arabidopsis." *Cell* 138 (4): 750–9. https://doi.org/10.1016/j.cell.2009.06.031.

Xue, Tao, Xuehuan Dai, Ruipu Wang, Junzhu Wang, Zhenhua Liu, and Fengning Xiang. 2017. "ARGONAUTE10 inhibits in vitro shoot regeneration via repression of miR165/166 in *Arabidopsis thaliana*." *Plant Cell Physiol* 58 (10): 1789–800. https://doi.org/10.1093/pcp/pcx117.

Yang, Ling, Lei Bian, Hai-long Shen, and Yu-hua Li. 2013a. "Somatic embryogenesis and plantlet regeneration from mature zygotic embryos of Manchurian ash (*Fraxinus mandshurica* Rupr.)." *Plant Cell Tissue Organ Cult* 15. https://doi.org/10.1007/s11240-013-0345-8.

Yang, Woorim, Myung-Hwan Choi, Bosl Noh, and Yoo-Sun Noh. 2020. "De novo shoot regeneration controlled by HEN1 and TCP3/4 in Arabidopsis." *Plant Cell Physiol* 61 (9): 1600–13. https://doi.org/10.1093/pcp/pcaa083.

Yang, Xiyan, Lichen Wang, Daojun Yuan, Keith Lindsey, and Xianlong Zhang. 2013b. "Small RNA and degradome sequencing reveal complex miRNA regulation during cotton somatic embryogenesis." *J Exp Bot* 64 (6): 1521–36. https://doi.org/10.1093/jxb/ert013.

Yang, Xiyan, and Xianlong Zhang. 2010. "Regulation of somatic embryogenesis in higher plants." *Crit Rev Plant Sci* 29 (1): 36–57. https://doi.org/10.1080/07352680903436291.

Yu, Niu, Qi-Wen Niu, Kian-Hong Ng, and Nam-Hai Chua. 2015. "The role of miR156/SPLs modules in Arabidopsis lateral root development." *Plant J* 83 (4): 673–85. https://doi.org/10.1111/tpj.12919.

Zhan, Junpeng, and Blake C. Meyers. 2023. "Plant small RNAs: their biogenesis, regulatory roles, and functions." *Annu Rev Plant Biol* 74 (1): 21–51. https://doi.org/10.1146/annurev-arplant-070122-035226

Zhang, Li-Feng, Yan-Ru Fan, Qian Lan, Li-Wang Qi, and Su-Ying Han. 2021. "Expression of the *SPL*-like gene *LaSPL9* in Japanese larch (*Larix leptolepis*) is regulated by miR156 during somatic embryogenesis." *Trees* 35 (5): 1727–37. https://doi.org/10.1007/s00468-020-02081-9. https://doi.org/10.1007/s00468-020-02081-9.

Zhang, Q. L., L. Y. Su, S. T. Zhang, X. P. Xu, X. H. Chen, X. Li, M. Q. Jiang, S. Q. Huang, Y. K. Chen, Z. H. Zhang, Z. X. Lai, and Y. L. Lin. 2020. "Analyses of microRNA166 gene structure, expression, and function during the early stage of somatic embryogenesis in *Dimocarpus longan* Lour." *Plant Physiol Biochem* 147: 205–14. https://doi.org/10.1016/j.plaphy.2019.12.014.

Zhang, Tian-Qi, Heng Lian, Hongbo Tang, Karel Dolezal, Chuan-Miao Zhou, Sha Yu, Juan-Hua Chen, Qi Chen, Hongtao Liu, Karin Ljung, and Jia-Wei Wang. 2015. "An intrinsic microRNA timer regulates progressive decline in shoot regenerative capacity in plants." *Plant Cell* 27 (2): 349–60. https://doi.org/10.1105/tpc.114.135186.

Zhang, Tian-Qi, Heng Lian, Chuan-Miao Zhou, Lin Xu, Yuling Jiao, and Jia-Wei Wang. 2017. "A two-step model for de novo activation of WUSCHEL during plant shoot regeneration." *Plant Cell* 29 (5): 1073–87. https://doi.org/10.1105/tpc.16.00863.

Zhu, Hongliang, Fuqu Hu, Ronghui Wang, Xin Zhou, Sing-Hoi Sze, Lisa Wen Liou, Ashley Barefoot, Martin Dickman, and Xiuren Zhang. 2011. "Arabidopsis Argonaute10 specifically sequesters miR166/165 to regulate shoot apical meristem development." *Cell* 145 (2): 242–56. https://doi.org/10.1016/j.cell.2011.03.024.

Zimmerman, J. Lynn. 1993. "Somatic embryogenesis: a model for early development in higher plants." *Plant Cell* 5 (10): 1411–23. https://doi.org/10.1105/tpc.5.10.1411.

6 Regeneration in Woody Plants
Challenges and Opportunities

Prem Prakash Jangid, Parth Ramesh Gharat, and Anurag Kumar

6.1 INTRODUCTION

There are around 600 species of timber plants worldwide, many of which have economic importance in the global market for plywood, roundwood, sawn wood and veneer. Many tree species are endangered because of logging, land conversion for agriculture, ineffective management strategies, and extensive deforestation that impede the natural regrowth of native forests. Multiple traditional propagation techniques are used for the regeneration of plant species, including seed propagation, cutting, layering, division, grafting, budding, and tissue culture. Tissue culture methods are particularly efficient in generating excellent planting stock as a result of their quick and exceptional multiplication rate. It involves the propagation of cells or tissues in a regulated and aseptic setting. *In vitro* micro-propagation of woody plants may be influenced by many variables, which might affect their culture or establishment. The elements influencing the process include the nature of the explants, the physiological state of the explant, the genetic makeup and age of the donor plant, the composition of the culture medium, the presence of plant growth hormones, the duration of light exposure, and the presence of antioxidants. The main challenges faced in the *in vitro* cultivation of woody plants pertain to the Juvenility-Maturity dilemma of the explant, the existence of endophytic fungus or microflora, contamination of cultures, discolouration of the medium due to leaching, vitrification or hyperhydricity, initiation of root formation in the shoots generated *in vitro* and the acclimatisation process. Plant cultivation and the growth of its various components in sterile conditions rely on the remarkable concept of totipotency. The core principle of plant tissue culture is rooted in the totipotent nature of plant cells and their ability to exhibit cellular plasticity—the capacity to divide, differentiate, and transform into various cell types.The ability of a single cell to undergo differentiation and develop into a whole plant closely mimic the parent plant. Tissue culture methods efficiently overcome difficulties in propagating plants that are difficult to replicate. This chapter analyses the most recent advancements in plant tissue culture methodologies, together with the limitations they face and possible remedies.

DOI: 10.1201/9781032702391-6

6.2 REGENERATION IN WOODY PLANTS

6.2.1 Macro-Propagation

A procedure that is used in a shed or out in the field is known as macro-propagation. There are two distinct sorts of procedures that fall under the category of macro-propagation. The first form of propagation involves the utilisation of vegetative sections of plants in order to raise new plants. When it comes to the reproduction of new plants, many vegetative characteristics of plants, such as shoots, leaves, roots, stems, buds, and subterranean sections, are used in a variety of ways. This includes cutting, layering, grafting and budding, which are the most frequent means of propagation. Each of these processes requires specialised knowledge and is carried out in a different manner in various plants (Njukwe et al., 2007).

Cutting: A cutting is a disengaged vegetative fragment of a plant that, upon separation and transplantation, has the capacity to restore the absent components and grow into a new plant itself. This technique of propagation is cost-effective and efficient. A multitude of identical plants may be generated from a limited number of parent plants. It does not need specialised expertise. The approach is referred to as the plant propagation technique that involves using certain parts of the plant for cutting, such as stem cutting, root cutting and leaf cutting.

Layering: It is a propagation technique that involves attaching one plant to another. This procedure involves the cultivation of roots on the covered section of the stem while it remains connected to the parent plant. Once the roots have emerged and developed, this section is detached from the parent plant and let to grow independently as a new plant on its own root stem. This kind of root stem is referred to as a 'layer'. There are many types of layering including simple layering, compound or serpentine stacking, trench layering, mound layering and air layering.

Grafting: The process of combining a part of two plants in such a way that they form a unit and operate as if they were coming from a single plant. When using this technique, the scion shoot is not separated from the mother plant until the union has taken place. After the scion and rootstock have successfully joined together, the scion is then gradually cut away from the mother plant to detach it from the plant. A container or a polythene bag is used to cultivate the rootstock so that the grafting process may be more convenient. To achieve effective graft unions in plants, which are notoriously difficult to do, this procedure is often used. The term 'approach grafting' refers to a certain kind of scion-added technique. In all, there are four distinct forms of it: (1) Grafting of the bark, (2) grafting of the cleft, (3) grafting of the side and (4) grafting of the veneer.

Budding: It is a specialized grafting technique where a fully matured bud from one plant is grafted onto the stem (rootstock) of another plant, enabling a successful union and the growth of a new plant. As a subset of grafting, budding is categorized into six distinct methods:

i. *T-Budding/Shield Budding*: This involves making a T-shaped incision on thin-barked rootstocks and inserting a shield-shaped bud, which is then secured with tape; it is commonly applied to plants like roses, apples, and grapes.
ii. *Patch Budding*: In this method, a square or rectangular bud is fitted into a matching incision on thick-barked rootstocks and tied with tape, making it ideal for species such as mangoes and jackfruits.
iii. *Chip Budding*: Suitable for dormant plants or those with low sap flow, this technique involves placing a rectangular bud with wood into a corresponding cut on the rootstock, commonly used for apples and grapes.
iv. *Ring Budding*: A ring-shaped bud is carefully fitted onto the rootstock without requiring a tie, making it popular for crops like plums, peaches, and mulberries.
v. *Modified Ring Budding*: An advanced version of ring budding, this method incorporates a vertical slit in the bud and involves tying with tape and applying wax for protection during rainy conditions.
vi. *Forkert Budding*: This method involves inserting a rectangular bud into a three-sided cut on the rootstock, which is then covered with bark and tied to prevent drying, typically used for plants like mangoes and cashews.

6.2.2 Micro-Propagation

Plant micro-propagation, or plant tissue culture, is a method that involves the isolation, sterilisation and incubation of cells, tissues or organs from selected plants in a sterile environment that promotes development. The objective is to produce a large quantity of plantlets. Cultivating tissues under controlled circumstances is a modern and innovative technique of vegetative growth. The process of propagation is highly specialised and allows for the production of a significant number of virus-free seedlings in a short period of time (Njukwe et al., 2007). Although there are several advantages to regenerating plants using plant tissue culture methods, there are several challenges and restrictions involved with the micro-propagation of woody plant species.

6.3 CHALLENGES IN REGENERATION OF WOODY PLANTS

6.3.1 Type and Physiological Status of Explants

In micro-propagation, it is preferable to use mother plants as the source of explants. This is because seeds, embryos, or seedlings do not possess sufficient information about the inherent characteristics of the plant, and it is challenging to assess the authenticity of the offspring (Debergh, 1987). Selection of explants in woody plants is often feasible at the later stages of their life cycle, after the juvenile phase has ended. However, establishing *in vitro* culture becomes challenging owing to the resistance of mature tissues. The origin of explants has been regarded as a crucial factor in the *in vitro* cultivation of woody plants. Not all explants possess the same regenerative capabilities. Different explants are expected to have distinct selection pressures.

(a) (b)

FIGURE 6.1 (a) *In vitro* shoot development from nodal explants of *Aegle marmelos* in woody plant medium (WPM +1.5 mg/L Kn). (b) *In vitro* developed shoot from nodal explants of *A. marmelos* in woody plant medium (WPM +1.5 mg/L Kn) for subculturing.

As a consequence, plants derived from various explants may exhibit distinct frequencies and spectrums of regeneration. The choice of explant is a crucial aspect in optimising the tissue culture process. The choice of explants, such as leaf, nodal, shoot tip, node, internode and root explants, has a major impact on the tissue culture process of plants (Feng et al., 2010, Girijashankar, 2012, Kumar et al., 2011). Cultivating woody plants via tissue culture has often presented challenges due to the cellulose and lignin composition of their cells and tissues. Consequently, primary shoots are generally seen to a short period when first burst take place in spring. The juvenility in mature tissues can be induced by extensive pruning (hedging) (Hatcher, 1959, Stoutemyer, 1944, Fazio, 1964, Mazalewski and Hackett, 1979) using epicormic shoot; and juvenile parts of mature plants (Stoutemyer et al., 1944, Fazio, 1964, Mazalewski and Hackett, 1979, Grace, 1939) using shoots developed from adventitious buds (Hacket, 1980); grafting mature scions on juvenile rootstocks (Doorenbos, 1954); sequential grafting (Muzik and Cruzado, 1958) treating mature plants with growth regulators (Paton et al., 1970); defoliating of mature trees (McCown and Sellmer, 1987) and use of various methods of *in vitro* propagation (Hamzazai, 2020) (Figure 6.1).

6.3.2 Sterilisation of Explants

The process of surface sterilisation of explants is a vital and critical stage in plant tissue culture. Exposing explants collected from the external environment to microbial contamination will result in the death of the explant. The exterior surface of the explant may be sterilised by using autoclave water and various chemical compounds such as ethanol, sodium hypochloride, bavistine, mercuric chloride and so on, to eliminate fungal and bacterial contamination. For culture, the explants were prepared by eliminating deceased tissue and cleansing them with tap water containing a small amount of detergent. This was then followed by 1–2 rinses with distilled water to eliminate the detergent. The $HgCl_2$ treatment varied in terms of the concentrations and time, depending on the kind of tissue used. For the shoot tips and nodal explants of woody plant species, a solution of $HgCl_2$ with a concentration of 0.1–1.0% (w/v) is applied for a duration of 2–5 minutes. Contamination in explants is controlled by testing pre-treatment or culture in the presence of antibiotics and antimycotics. Bavistin is a systemic fungicide that is used as a 0.2% (w/v) solution for a soaking treatment lasting 30 minutes. Streptomycin and gentamycin are introduced into the medium at a conçentration of 100 mg/L after filter sterilisation using a 0.22-micron membrane (millipore), just before to the media being autoclaved and solidified. The use of 0.2% (w/v) Bavistin in the medium proved to be efficacious in lowering the contamination to below 60% and in retarding the onset of contamination.

6.3.3 Culture Medium

The culture medium plays a vital role in the micro-propagation of woody plants due to the distinct nutritional needs of woody plants compared to herbaceous plants. Plant tissue culture media comprises essential macronutrients, micronutrients, vitamins, organic components, plant growth regulators, carbon sources and gelling agents. These components are necessary for the optimal growth and development of plants in a solid medium. The development of many woody species is hindered by high ionic strength, while low salt formulations do not provide sufficient support for shoot growth (Andreea et al., 2009). The pH level of the base medium is crucial as it has an impact on both plant development and the functioning of growth regulators. The formulation of the fundamental medium and its integration with phytohormones and nitrogen sources significantly influence the initiation of the explanatory response. The commonly recognised basal media for tissue culture include Murashige and Skoog medium (MS), White medium (WM), Vacant Went medium (VW), Nitsch and Nitsch medium, Schenk and Hildebrandt medium, Woody Plant Medium (WPM) and N6 medium (Van Nieuwkerk et al., 1986). In general, the MS medium is often used for herbaceous plants, but the WPM and Schenk and Hildebrandt medium are specifically designed for the woody plants.

6.3.4 Plant Growth Regulators

Plant growth regulators (PGRs) are crucial in influencing the developmental pathways of plant cells and tissues in the culture medium. Auxins, cytokinins, and gibberellin are the main growth regulators used in practice. The selection and dosage of hormones used are essentially determined by the specific plant species, the cultured tissue or organ, and the experimental objectives. Auxins and cytokinins are the predominant growth regulators used in plant tissue culture, and their quantity defines the specific kind of culture that is established or regenerated. Elevated levels of auxin often facilitate the development of roots, while increased levels of cytokinins stimulate the regrowth of shoots. A precise equilibrium between auxins and cytokinins results in the formation of an undifferentiated cell cluster referred to as a callus. The usage of PGRs in woody plant tissue culture varies significantly due to their susceptibility to factors such as type, composition, concentration, regeneration, and subculture stage. BAP and NAA are the top options in woody plant tissue culture. The process of callus induction, proliferation, and regeneration in different plant species has been investigated by using a combination of BAP and NAA (Devi et al., 2022). Various data indicate that the inclusion of BAP together with 2, 4-phenoxy acetic acid

(2.4 D) might result in necrosis in explants. Thidiazuron is a fascinating cytokinin-like growth regulator that has been just introduced. It has significant potential for the micro-propagation of several arboreal species (Hammerschlag, 1986).

6.3.5 Seasonal Effect

Within the tissue culture regeneration system, the season is a critical component that needs meticulous consideration. The season in which the explant collection takes place is of great importance, as mentioned before (Fernandez-Garcia et al., 2008). The success rate of transplants reaches its zenith from January to March. The mishandling of explants increases the likelihood of microbial infection. Explants collected from May to November showed absolute absence of fungal infection, irrespective of sterilisation, antifungal treatment, or antibiotics. The pollution rates reached their peak in April and March, with a recorded rate of 45%. Buds could only live without contamination during the dry season.

6.3.6 Hyperhydricity or Vitrification

Hyperhydricity, formerly referred to as vitrification, encompasses several physiological dysfunctions resulting from aberrant water conditions inside a tissue culture container. This issue is significant and has the potential to impact the process of shoot multiplication and the overall strength of the culture (Dewir et al., 2006). Hyperhydricity may result in hypertrophy, a deficit of chlorophyll, insufficient lignification of cell walls, the presence of wide intercellular gaps in both spongy mesophyll and palisade cells and non-functional stomata in leaves that seem normal. Additionally, it can lead to poor survival and roots when transplanted to a greenhouse. Despite the many publications in this field, there is still no agreement on the clear and definitive processes of hyperhydricity. According to the prevailing consensus, the development of organisms under stressful settings results in the formation of reactive oxygen species (ROS) (Balen et al., 2009, Jaspers et al., 2010, Ziv et al., 1987). Hyperhydricity results in the production of O_2, H_2O_2, and OH primarily in apoplasts, chloroplasts, mitochondria, and the cytosol (Ozyigit, 2008). Additionally, it was discovered that H_2O_2 is specifically located inside the cell walls and intercellular gaps of hyperhydric plants (Balen et al., 2009).

There are several factors that can be utilised to eliminate the vitrification problem. For instance, $AgNO_3$ and Ag nanoparticles are commonly employed as additives in the nutrient medium to decrease the occurrence of hyperhydricity. This is because silver ions can hinder the transmission of the ethylene signal by binding to ethylene receptors. Muneer et al. (2006) found that red and blue LEDs effectively decreased hyperhydricity to a manageable extent. The impact of a gelling compound on hyperhydricity in plants is thought to be contingent upon its specific kind and brand (Renu et al., 2021). The most effective approach for mitigating hyperhydricity in *in vitro* culture is to decrease the concentration of cytokinins in the media. A reduction in hyperhydricity is seen when using a medium with a lower mineral salt concentration (Weatherhead et al., 1979). The process of vitrification may be regulated by chilling the lower part of the container used for tissue culture, within the temperature range of 40°C–42°C in shoot cultures of *Dianthus caryophyllus*.. Implementing any of the aforementioned methods to mitigate hyperhydricity may not always provide satisfactory results. Various plant species may need distinct substrates and cultivation parameters. Regarding this matter, it is necessary to do thorough and individualised studies on the hyperhydricity of each species. Furthermore, the methods used to eliminate hyperhydricity should be multifaceted and tailored to the distinctive characteristics of each species.

6.3.7 Browning of Culture Medium due to Leaching

The presence of moderate browning in the cultivation of woody plants is mostly attributed to the substantial production of phenols, especially in mature explants such as nodal segments and shoot apices. This behaviour eventually leads to the demise of plants (Bharadwaj and Ramawat, 1993). Phenolic substance leaches from the cut surface of the explant and oxidises later results in turning of media brown. Browning of tissue process is caused by the oxidation of tannin and polyphenols and the formation of guinones which are highly reactive and toxic to the tissue. Antioxidants remove free radical intermediates and inhibit other oxidation reaction. Several oxidases such as monophenolase (tryosinase), polyphenol (catecholoxidase) oxidised the hydroxy group result in the formation of guinone and water (Fridborg et al., 1978). This wounding further facilitates the release of various enzymes such as polyphenol oxidase, superoxide dismutase, and peroxidase in the wounded plant part. These enzymes come to the rescue of the plant by catalysing various reactions to eliminate reactive oxygen and heal the plant part. One of these processes leads to the synthesis of 'melanin', a dark pigment, which causes the medium and explants to become brown. The formation of significant amounts of melanin resulting from the oxidation of phenolic chemicals may cause a decrease in plant development, slower rates of regeneration or recalcitrance and ultimately lead to tissue or plant death. In order to control browning of medium and explants *in vitro*, many researchers have tried to incorporating external substances in the tissue culture media. The solutions for leaching problems are as follows:

- *Absorbants*: incorporation of nonspecific absorbents like activated charcoal, ascorbic acid and polyvinylpyrrolidone in to the tissue culture medium plays an important role in irreversible adsorption of inhibitory compounds from the culture medium. Apart from this these reduces the toxic metabolites, phenolic exudation from culture media (Thomas, 2008, Ishtiaq Ahmad et al., 2013, Leifert et al., 2001, Ulrich et al., 2008).
- *Antioxidants*: pre-soaking of explants in an antioxidant solution, incorporation of antioxidants into the medium and it is also used to control the leaching problem in the culture medium. By scavenging oxygen radicals generated when plant tissue is injured, these antioxidants shield the cells from oxidative damage. Some common antioxidants include ascorbic acid, citric acid, polyvidone, cysteine, and silver nitrate.
- Activated charcoal is effective in trapping toxins and inhibitory compounds, preventing their absorption by plants. Activated charcoal may be added to

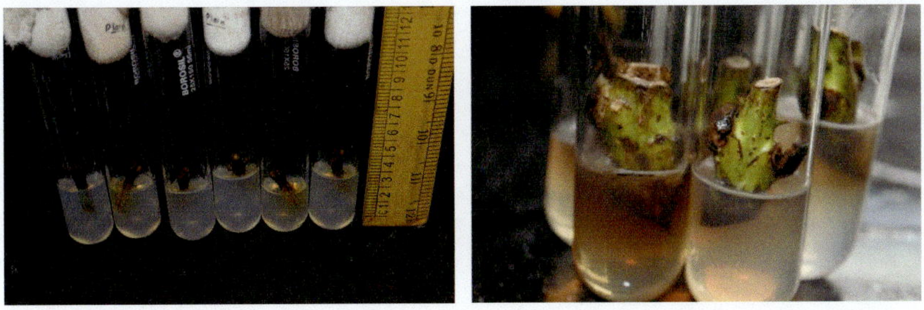

(a) (b)

FIGURE 6.2 (a, b) Leaching from *in vitro* grown nodal stem explants in culture medium of *Gmelina arborea*.

the culture medium to lessen the build-up of oxidised phenolic compounds. Additionally, it enhances tissue growth and development.

- *Dark Culturing*: Dark periods and regular subculturing of explants are necessary for the culture incubation process.
- Light has been found to enhance the synthesis of phenolic compounds. Therefore, it is recommended to culture the explants in a dark environment for a duration of 72 hours to ten days in order to effectively minimise the occurrence of browning. The length of time for culturing in the dark varies depending on the specific species being studied.
- *Polyphenol Inhibitors*: Specific compounds or chemicals possess the ability to impede the production of phenolic compounds in tissue culture circumstances. The addition of these inhibitors can result in substantial reductions in visual tissue discolouration and a decrease in total phenol content, which in turn leads to a substantial increase in tissue growth and development. Some examples include sodium chloride, sulphur dioxide, and 2-aminoindane-2-phosphonic acid.
- *Subculturing*: Regularly culturing explants/shoots to prevent discolouration. Regular subculturing is essential for maintaining optimal growth conditions and preventing the accumulation of phenolic compounds. This ultimately results in enhanced cultural growth and development (Figure 6.2).

6.3.8 Endophytic Fungus and Microflora

Endophytic bacteria and fungus are generally found in plant *in vitro* cultures, both in commercial as well as in scientific laboratories which affect the *in vitro* propagation of woody plant species (Leifert et al., 2001, Parkinson et al., 1996). The plant–endophyte relationship can be challenging for plants that are propagated in tissue culture (Crovadore et al., 2012). For *in vitro* culture, the plants are surface sterilised to remove any external bacteria and fungi before being grown in sterile sucrose-rich culture conditions. Surface sterilisation methods are effective at reducing or eliminating surface contaminants, but they usually do not affect endogenous microbes because the sterilisation methods do not penetrate deeply into the plant tissue.

FIGURE 6.3 Fungal and bacterial contamination in *in vitro* culture of nodal explants.

The agents used for the surface sterilisation of plant tissues include sodium hypochlorite, calcium hypochlorite, ethanol or isopropanol, hydrogen peroxide, chlorine gas, sodium dichloroisocyanurate, isothiazolone-based biocide, antibiotics and mercuric chloride (Bele et al., 2012, Mo et al., 2008). Epiphytic microbes in the explant were removed by surface sterilisation, but the occurrence of endophytic microbes in the explant leads to contamination in the culture. The use of antibiotics plays an essential role in the elimination and inhibition of microbial contamination by adding appropriate amount of antibiotics like streptomycin, tetracycline, gentamicin and cefotaxime in the medium may eradicate the contamination in culture. Benomyl and streptomycin were added in the medium provides better culture of nodal explant (Figure 6.3).

6.3.9 Establishment of Roots in *In Vitro* Grown Shoots

Difficulty with rooting is a major problem in the final stage of shoot organogenesis in woody plants (Thimann et al., 1934, Chalupa, 1984) and is often encountered in the process of somatic embryogenesis (Mato and Vieitez, 1986) Rooting in the *in vitro* produced shoots is another problem in many tree species. *Ex vitro* rooting has been found highly successful in some species. Auxin's involvement in adventitious root formation has been established for a long time. Thimann and Went demonstrated the significance of indole-acetic acid (IAA) in the root generation process as early as 1934. Garcia-Gomez et al. (1994) found that auxins are crucial in the process of inducing adventitious roots in avocado. Garcia-Gomez et al. (1994) discovered that the number of roots produced in avocado plants can be enhanced by the addition of indole-3-butyric acid (IBA) to the culture medium. They observed that while juvenile avocado shoots naturally produce roots even without the presence of external auxin, the inclusion of IBA resulted in an increased number of roots per plant. The addition of exogenous auxin is a prerequisite for root induction in woody plants with difficult in root generation, i.e. oak (Hammatt and Ridout, 1992) chestnut (Hausman et al., 1994), ash (Ripetti et al., 1994) and poplar (Chandra et al., 2010). However, IAA is

FIGURE 6.4 Root establishment in *in vitro* grown shoot.

considered to be the major auxin for initiation of rooting, but IBA is also used widely. Ripetti et al. (1994) also reported a beneficial effect of a dark pre-treatment during root initiation in walnut (Figure 6.4).

6.3.10 Acclimatisation

An important constraint in the widespread use of micro-propagation technology is the significant death rate seen in plants cultivated *in vitro* when they are transferred from the laboratory to the field. When micro-propagated plants are moved to circumstances outside of a controlled environment, such as changes in temperature, light intensity and water availability, they need acclimatisation to successfully establish and survive as plantlets (Birchfield and McIntosh 2020). It is crucial to investigate and comprehend the characteristics that govern the effective adaptation of tissue grown plants. Additionally, it is essential to acknowledge and grasp the distinctions between an *in vitro* setting and a greenhouse or field environment. The applicant has recognised the significance of acclimatisation for micro-propagated plants and has endeavoured to collect existing knowledge on the morphology and anatomy of plants cultivated *in vitro, ex vitro*, or in the wild. Transferring shoots or plantlets from a controlled environment to a greenhouse may lead to fast dehydration, wilting, and even death if proper measures are not followed to adapt them to the new surroundings (Figure 6.5).

(a) (b)

FIGURE 6.5 Acclimatisation of *in vitro* grown plantlets in hardening room and in open environment.

6.4 OPPORTUNITIES IN REGENERATION OF WOODY PLANTS

Plant products consist of a variety of compounds that are crucial for plant metabolism and physiology. The genetic composition of plants has a significant impact on the phytochemical content they possess. Phytoconstituents refer to a diverse range of bioactive chemicals derived from plants, such as terpenoids, alkaloids and flavonoids. Plants depend on phytochemicals for their existence; however, these compounds also possess several medicinal applications and play a vital role in the functioning of plants. Due to their medicinal properties, these compounds may provide protection against a broad spectrum of diseases. Cancer, diabetes and coronary heart disease are among the illnesses that might potentially derive therapeutic benefits from their use as an active pharmaceutical ingredient. They possess a multitude of applications, including serving as dietary supplements, providing natural pigments, and fulfilling other purposes. Consequently, there is a strong incentive to capitalise on the production of these metabolites due to their significant economic value.

A plant's genetic makeup influences both the synthesis and control of these phytochemicals. The elucidation of biosynthetic pathways and regulatory elements has enabled the successful integration of whole pathways into microbial systems, as well as the manipulation of plant metabolism to introduce novel capabilities. One approach to accomplish this objective is to use metabolic engineering in plants, which involves increasing the production of desired molecules or reducing the production of undesired ones. Application of synthetic biology techniques to microorganisms has the capacity to profoundly transform the pharmaceutical and nutraceutical industries (Parray et al., 2019).

6.5 METABOLIC ENGINEERING TECHNOLOGIES

Modifying natural product biosynthetic pathways can occur in plants. Generating a knock-out line or introducing a new gene by knock-in process has been proven to regulate the production and also in confirming the function of that particular gene or protein. There are several instances of transgenic plants, and their creation has developed to fulfil various purposes (Hoshino et al., 2019). Recently, a method has been developed to create a transgenic morning glory plant that produces AmAS1 with Am4OCGT. This leads to the build-up of aureusidin glucosides, resulting in a variety of the flower with a distinct hue. The yellow colour of flower petals is partly due to the accumulation of molecules belonging to the aurone subclass of flavonoids (Alquézar et al., 2021). Additional endeavours have been made to genetically manipulate the build-up of plant secondary metabolites, which are recognised for their beneficial effects on health, in essential crops like maize or wheat. Conversely, many research have shown encouraging outcomes *via* the use of transgenic plants as a means of obtaining pharmaceutical or nutraceutical substances. This process entails the gathering of certain secondary metabolites in microbial systems. This technique has shown its benefits in some instances (Parray et al., 2019).

- The technique of metabolic engineering in fruit trees, which includes enhancing nutritional content and optimising postharvest metabolic processes, significantly enhances the value of the final product that may be consumed. Furthermore, the designed product enhances global human well-being by incorporating flavonoids, polyphenols and carotenoids into fruits. For instance, in transient transgenic apple Calli, downregulating the expression of MdMYB44, which is a regulator of acidity in apples, resulted in increased build-up of malate, a crucial component which provides flavour (Hoshino et al., 2019). The build-up of sugar in apples is caused by the suppression of the important enzyme aldose-6-phosphate reductase, which stimulates the manufacture of sorbitol from glucose-6-phosphate (Alquézar et al., 2021). In another instance, a genetically modified apple plant was created to consistently produce higher levels of anthocyanin and exhibit a red colour in its fruit flesh. This was achieved by introducing the MdWRKY11 transcription factor, which regulates the production of anthocyanin (Li et al., 2018, Liu et al., 2019).

- CRISPR-Cas genetic engineering in plants has great potential for promoting environmental sustainability, addressing food security concerns and tackling agricultural difficulties caused by climate change, both biotic and abiotic in nature. CRISPR-Cas technology has been well noticed for its use, particularly in the field of plant genome editing. Nanomaterials have the potential to tackle key challenges related to CRISPR genome editing in plants, such as cargo delivery, species independence, germline transformation and gene editing efficiency.
- Biotechnological methods, such as tissue culture techniques including callus culture, hairy root culture, protoplast culture, and micro-propagation procedures, have shown noteworthy outcomes in the production of secondary metabolites (Pons et al., 2014). Novel approaches based on plant tissue culture technologies provide benefits such as easier manipulation of sources and enhanced extraction of secondary components in terms of both quality and quantity. Plant tissue culture techniques, such as callus culture and *in vitro* propagation, allow the generation of secondary compounds at a faster rate and higher quantity while preserving the negative impact on the environment (Ozyigit et al., 2023, Beigmohamadi et al., 2019). Recombinant DNA technology enables the transfer of certain genes, such as GUS (a reporter gene), NPT II (a marker gene), dehE and dehD (genes for herbicide tolerance), across different species (Kong et al., 2019, Mohamed et al., 2016). Recombinant DNA techniques have shown promising results for increasing the production of some secondary chemicals through modification of the secondary metabolite pathways.

6.6 CONCLUSION

Ultimately, the process of regenerating woody plants encompasses a range of techniques, including macro-propagation using cutting, layering, grafting, budding and micro-propagation. These techniques are essential for the spread and nurturing of attractive plant kinds. Nevertheless, obstacles endure throughout the process of regeneration. Challenges arise during regeneration of woody plants through plant tissue culture from several factors such as the nature and physiological condition of the plant tissue used, the sterilisation process for the tissue, the composition of the growth medium, the presence of PGRs, the influence of seasonal variations, the occurrence of vitrification, the contamination by endophytic fungi and microflora and the discolouration of the medium caused by leaching. The success of regeneration also relies on the development of roots in shoots produced in a controlled environment and the adaptation of regenerated plants to the outside conditions. To overcome these problems, one must possess a comprehensive comprehension of the complex interaction among many components and engage in ongoing research to enhance the efficiency of regeneration treatments. The sustainable regeneration of woody plants may be achieved by resolving the obstacles *via* developments in biotechnology and plant tissue culture techniques. This will help conserve and propagate significant genetic resources for future generations. Efficient methods of regrowth are crucial not only for the cultivation of plants for business purposes but also for the process of ecological renewal and the protection of biodiversity. If researchers modify key genes involved in growth, stress resistance and reproductive activity, they have the ability to accelerate regeneration, increase yield and improve tolerance to changing environmental conditions. Regarding the use of genetic engineering in woody plants, it is essential to analyse ethical concerns, regulatory structures and the possible ecological consequences. The development of genetically modified woody plants with traits such as enhanced insect resistance, greater climate change resilience and accelerated growth rates offers several prospects. Potential consequences of these genetically engineered plants include enhanced timber production, optimised carbon sequestration and ecosystem restoration in areas that have suffered degradation. Plant tissue culture with genetic engineering might potentially aid in the conservation of endangered woody plant species by facilitating their multiplication and reintroduction into their native habitats. Hence, to address the challenges linked to the regrowth of woody plants, a comprehensive strategy including several disciplines is necessary, whereby genetic engineering assumes a crucial function in the creation of novel prospects. Genetic engineering methods that are responsible and sustainable have the capacity to significantly contribute to the conservation and use of woody plants, benefiting both ecosystems and human societies. This is due to the rapid progress of technology and the expanding knowledge in the field of plant biology. In order to fully harness the capabilities of genetic engineering for woody plant regeneration, it will be imperative to engage in continuous research and collaborate with other scientists.

REFERENCES

Alquézar, Berta, Haroldo Xavier Linhares Volpe, Rodrigo Facchini Magnani, Marcelo Pedreira De Miranda, Mateus Almeida Santos, Viviani Vieira Marques, Márcia Rodrigues de Almeida et al. "Engineered orange ectopically expressing the Arabidopsis β-caryophyllene synthase is not attractive to *Diaphorina citri*, the vector of the bacterial pathogen associated to huanglongbing." *Frontiers in Plant Science* 12 (2021): 641457.

Andreea, Nistor, G. H. Campeanu, N. Chriu, and D. Kracsonyi. "Effect of auxin and cytokinin on callus induction in potato (*Solanum tuberosum* L.) explants." *Agricultura Stiintasipractica* 12 (2009): 47–50.

Balen, Biljana, Mirta Tkalec, Dubravko Pavoković, Branka Pevalek-Kozlina, and Marijana Krsnik-Rasol. "Growth conditions in in vitro culture can induce oxidative stress in *Mammillaria gracilis* tissues." *Journal of Plant Growth Regulation* 28 (2009): 36–45.

Bele, D., Tripathi, M.K., Tiwari, G., Baghel, B.S., and Tiwari, S. "Microcloning of sandalwood (Santalum album Linn.) from cultured leaf discs." *Journal of Agricultural Technology* 8, no. 2 (2012): 571–583.

Beigmohamadi, Mina, Ali Movafeghi, Ali Sharafi, Samineh Jafari, and Hossein Danafar. "Cell suspension culture of *Plumbago europaea* L. towards production of plumbagin." *Iranian Journal of Biotechnology* 17, no. 2 (2019): e2169.

Bhardwaj, L. A. T. A., and K. G. Ramawat. "Effect of anti-oxidants and adsorbents on tissue browning associated metabolism in Cocculus pendulus callus cultures." *Indian Journal of Experimental Biology* 31, no. 8 (1993): 715–718.

Birchfield, Aaron S., and Cecilia A. McIntosh. "Metabolic engineering and synthetic biology of plant natural products–A minireview." *Current Plant Biology* 24 (2020): 100163.

Chalupa, V. "*In vitro* propagation of oak (*Quereus robur* L.) and linden (*Tilia cordata* Mill.)." *Biologia Plantarum* 26, no. 5 (1984): 374–377.

Chandra, Sheela, Rajib Bandopadhyay, Vijay Kumar, and Ramesh Chandra. "Acclimatization of tissue cultured plantlets: From laboratory to land." *Biotechnology Letters* 32 (2010): 1199–1205.

Crovadore, Julien, Michel Schalk, and François Lefort. "Selection and mass production of *Santalum album* L. calli for induction of sesquiterpenes." *Biotechnology & Biotechnological Equipment* 26, no. 2 (2012): 2870–2874.

Debergh, P. C. "Improving micropropagation." *IAptC Newsletter* 51, no. 1 (1987): 2–10.

Devi, Sunita, and Neelam R. Yadav. "Role of explants collecting seasons and sterilization process on in vitro regeneration of guava: A technology hurdle." *The Pharma Innovation Journal* 11, no. 5 (2022): 1244–1248.

Dewir, Y. H., D. Chakrabarty, M. B. Ali, E. J. Hahn, and K. Y. Paek. "Lipid peroxidation and antioxidant enzyme activities of Euphorbia milliihyperhydric shoots." *Environmental and Experimental Botany* 58, no. 1–3 (2006): 93–99. doi: 10.1016/j.envexpbot.2005.06.019.

Doorenbos, J. "Rejuvenation of *Hedera helix* in graft combinations." *Proceedings. Koninklijke Nederlandse Akademie van Wetenschappen, Series C* 57 (1954): 99–102.

Fazio, S. "Propagating eucalyptus from cuttings." *Combined Proceedings International Plant Propagators' Society* 14 (1964): 288–289.

Feng, Jian-Can, X. M. Yu, X. L. Shang, J. D. Li, and Y. X. Wu. "Factors in-fluencing efficiency of shoot regeneration in *Ziziphus jujuba* Mill. 'Huizao'." *Plant Cell, Tissue and Organ Culture (PCTOC)* 101 (2010): 111–117.

Fernandez-Garcia, N., A. Piqueras, and E. Olmos. "Sub-cellular location of H_2O_2, peroxidases and pectin epitopes in control and hyperhydric shoots of carnation." *Environmental and experimental botany* 62, no. 2 (2008): 168–175.

Fridborg, Gunnar, Marianne Pedersen, Lars-Erik Landström, and Tage Eriksson. "The effect of activated charcoal on tissue cultures: Adsorption of metabolites inhibiting morphogenesis." *Physiologia Plantarum* 43, no. 2 (1978): 104–106.

García-Gómez, M.L., Sánchez-Romero, C., Barceló-Munoz, A., Heredia, A., Pliego-Alfaro, F. "Levels of endogenous indole-3-acetic acid and indole-3-acetyl-aspartic acid during adventitious rooting in avocado microcuttings." *Journal of Experimental Botany* 45, no. 6 (June 1994): 865–870. doi: 10.1093/jxb/45.6.865.

Girijashankar, V. "In vitro regeneration of *Eucalyptus camaldulensis*." *Physiology and Molecular Biology of Plants* 18, no. 1 (2012): 79–87.

Grace, N. H. "Vegetative propagation of conifers: I. Rooting of cuttings taken from the upper and lower regions of a Norway spruce tree." *Canadian Journal of Research* 17 (1939): 178–180.

Hackett, Wesley P. "Control of phase change in woody plants." In Little, C. H. A. (ed), *Control of Shoot Growth in Trees. Proceedings of the International Union for Research Organization Working Parties on Xylem Physiology and Shoot Growth Physiology.* Fredericton, Canada: University of New Brunswick, 1980.

Hammatt, N., and M. S. Ridout. "Micropropagation of common ash (*Fraxinus excelsior*)." *Plant Cell, Tissue and Organ Culture (PCTOC)* 31 (1992): 67–74.

Hammerschlag, F. A. "Temperate fruits and nuts." In Zimmerman, R. H., Griesbach, R. J., Hammerschlag, F. A., and Lawson, R. H. (eds), *Tissue Culture as a Plant Production System for Horticultural Crops*, pp. 221–236. Dordrecht: Martinus Nijhoff Publishers, 1986.

Hamzazai, Aasiya. "Factors Shaping Endophyte Communities Associated with Selected Cultivated Plants in Arizona." PhD diss., The University of Arizona, 2020.

Hatcher, E. S. J. "The propagation of rootstocks from stem cuttings." *Annals of Applied Biology* 47 (1959): 635–639.

Hausman, Jean-François, Claire Gevers, and Thomas Gaspar. "Involvement of putrescine in the inductive rooting phase of poplar shoots raised in vitro." *Physiologia Plantarum* 92, no. 2 (1994): 201–206.

Hoshino, Atsushi, Takayuki Mizuno, Keiichi Shimizu, Shoko Mori, Sachiko Fukada-Tanaka, Kazuhiko Furukawa, Kanako Ishiguro, Yoshikazu Tanaka, and Shigeru Iida. "Generation of yellow flowers of the Japanese morning glory by engineering its flavonoid biosynthetic pathway toward aurones." *Plant and Cell Physiology* 60, no. 8 (2019): 1871–1879.

Ishtiaq Ahmad, Ishtiaq Ahmad, Tanveer Hussain Tanveer Hussain, Irfan Ashraf Irfan Ashraf, Muhammad Nafees Muhammad Nafees, Maryam Maryam, Muhammad Rafay Muhammad Rafay, and Muhammad Iqbal Muhammad Iqbal. "Lethal effects of secondary metabolites on plant tissue culture." *American-Eurasian Journal of Agricultural & Environmental* 13 (2013): 539–547.

Jaspers, Pinja, and Jaakko Kangasjärvi. "Reactive oxygen species in abiotic stress signaling." *Physiologia Plantarum* 138, no. 4 (2010): 405–413.

Kong, Chui-Hua, Tran Dang Xuan, Tran Dang Khanh, Hoang-Dung Tran, and Nguyen Thanh Trung. "Allelochemicals and signaling chemicals in plants." *Molecules* 24, no. 15 (2019): 2737.

Kumar, Nitish, K. G. Vijay Anand, and Muppala P. Reddy. "Plant regeneration of non-toxic Jatropha curcas—impacts of plant growth regulators, source and type of explants." *Journal of Plant Biochemistry and Biotechnology* 20 (2011): 125–133.

Leifert, Cassells, and A. C. Cassells. "Microbial hazards in plant tissue and cell cultures." *In Vitro Cellular & Developmental Biology-Plant* 37 (2001): 133–138.

Li, Mingjun, Pengmin Li, Fengwang Ma, Abhaya M. Dandekar, and Lailiang Cheng. "Sugar metabolism and accumulation in the fruit of transgenic apple trees with decreased sorbitol synthesis." *Horticulture Research* 5 (2018): 60.

Liu, Wenjun, Yicheng Wang, Lei Yu, Huiyan Jiang, Zhangwen Guo, Haifeng Xu, Shenghui Jiang et al. "MdWRKY11 participates in anthocyanin accumulation in red-fleshed apples by affecting MYB transcription factors and the photoresponse factor MdHY5." *Journal of Agricultural and Food Chemistry* 67, no. 32 (2019): 8783–8793.

Mato, M. C., and A. M. Vieitez. "Changes in auxin protectors and IAA oxidases during the rooting of chestnut shoots in vitro." *Physiologia Plantarum* 66, no. 3 (1986): 491–494.

Mazalewski, R. L., and W. P. Hackett. "Cutting propagation of *Eucalyptus ficifolia* using cytokinin-induced basal trunk shoots." In *Combined Proceedings International Plant Propagators' Society*, vol. 29, 1979.

McCown, B. H., and Sellmer, J. C. "General media and vessels suitable for woody plant culture." In *Cell and Tissue Culture in Forestry: General Principles and Biotechnology*, pp. 4–16. Dordrecht: Springer Netherlands, 1987.

Mo, X. L., Q. Q. Zeng, W. F. Qiu, and Y. Z. Chen. "Study on somatic embryogenesis from sandalwood and plantlet regeneration." *Food Drug* 10 (2008): 35–37.

Mohamed, E., F. A. Rahiman, R. A. Wahab, C. R. Zain, M. A. Javed, and F. Huyop. "A plant transformation vector containing the gene dehD for the development of cultivars resistant to monochloroacetic acid." *JAPS: Journal of Animal & Plant Sciences* 26, no. 4 (2016): 1133–1139.

Muzik, T. J., and H. J. Cruzado. "Transmission of juvenile rooting ability from seedlings to adults of *Hevea brasiliensis*." *Nature* 181 (1958): 1288.

Njukwe, E., A. Tenkouano, D. Amah, K. Sadik, M. Perez, M. Nyine, and T. Dubois. *Training Manual. Macro-Propagation of Banana and Plantain*. Ibadan: IITA, 2007.

Ozyigit, Ibrahim Ilker, Ilhan Dogan, Asli Hocaoglu-Ozyigit, Bestenur Yalcin, Aysegul Erdogan, Ibrahim Ertugrul Yalcin, Evren Cabi, and Yilmaz Kaya. "Production of secondary metabolites using tissue culture-based biotechnological applications." *Frontiers in Plant Science* 14 (2023): 1132555.

Ozyigit, Ibrahim Ilker. "Phenolic changes during in vitro organogenesis of cotton (*Gossypium hirsutum* L.) shoot tips." *African Journal of Biotechnology* 7, no. 8 (2008): 1145–1150.

Parkinson, M., M. Prendergast, and A. J. Sayegh. "Sterilisation of explants and cultures with sodium dichloroisocyanurate." *Plant Growth Regulation* 20 (1996): 61–66.

Parray, Javid Ahmad, Mohammad Yaseen Mir, Nowsheen Shameem, Javid Ahmad Parray, Mohammad Yaseen Mir, and Nowsheen Shameem. "Plant genetic engineering and GM crops: Merits and demerits." In *Sustainable Agriculture: Biotechniques in Plant Biology*, pp. 155–229, 2019.

Paton, D. M., R. R. Willing, W. Nicholls, and L. D. Pryor. "Rooting of stem cuttings of *Eucalyptus*: A rooting inhibitor in adult tissue." *Australian Journal of Botany* 18, no. 2 (1970): 175–183.

Pons, Elsa, Berta Alquézar, Ana Rodríguez, Patricia Martorell, Salvador Genovés, Daniel Ramón, María Jesús Rodrigo, Lorenzo Zacarías, and Leandro Pena. "Metabolic engineering of β-carotene in orange fruit increases its in vivo antioxidant properties." *Plant Biotechnology Journal* 12, no. 1 (2014): 17–27.

Renu, Chandola, and Upadhyaya Anshulika. "Control of browning during seed germination of *Meizotropis pellita*: A rare and an endangered plant." *Research Journal of Biotechnology* 16 (2021): 8.

Ripetti, V., C. L. Kevers, and Th Gaspar. "Two successive media for the rooting of walnut shoots in vitro. Changes in peroxidase activity and in ethylene production." *Advances in Horticultural Science* 8, no. 1 (1994): 29–32.

Stoutemyer, V. T., F. L. O'Rourke, and Wilmer W. Steiner. "Some observations on the vegetative propagation of honey locust." *Journal of Forestry* 42, no. 1 (1944): 32–36.

Thimann, Kenneth V., and Frits Warmolt Went. "On the chemical nature of the root-forming hormone." *Proceedings. Koninklijke Nederlandse Akademie van Wetenschappen* 37 (1934): 456–459.

Thomas, T. Dennis. "The role of activated charcoal in plant tissue culture." *Biotechnology Advances* 26, no. 6 (2008): 618–631.

Ulrich, Kristina, Thomas Stauber, and Dietrich Ewald. "*Paenibacillus*—a predominant endophytic bacterium colonising tissue cultures of woody plants." *Plant Cell, Tissue and Organ Culture* 93, no. 3 (2008): 347–351.

Van Nieuwkerk, J. P., Richard H. Zimmerman, and Ingrid Fordham. "Thidiazuron stimulation of apple shoot proliferation in vitro." *HortScience* 21, no. 3 (1986): 516–518.

Weatherhead, M. A., J. Burdon, and G. G. Henshaw. "Effect of activated charcoal as an additive to plant tissue culture media. Part 2." *Pflanzenphysiologie* 94 (1979): 399–406.

Ziv, Meira, Amnon Schwartz, and Devorah Fleminger. "Malfunctioning stomata in vitreous leaves of carnation (*Dianthus caryophyllus*) plants propagated in vitro; implications for hardening." *Plant Science* 52, no. 1–2 (1987): 127–134.

7 Regeneration of Endangered Medicinal Plants

Kunal Ranjan, Supriyo Basak, Rajani Sharma, and Kunal Kumar

7.1 INTRODUCTION

Plants possess the remarkable ability to produce a wide range of metabolites that are highly sought after for their pharmaceutical, flavor-enhancing, coloring, biopesticide, and food additive properties. These compounds, including alkaloids, essential oils, resins, and tannins, are extracted from plants with medicinal properties and serve as vital building blocks for essential medicines and drugs (Hamilton, 2004; Simmonds, 2005). More than 10% of plant species that constitute over 50,000 are utilized in the global production of pharmaceuticals and health products. However, the distribution of these valuable medicinal plants is not uniform throughout the world (Srujana, Babu, and Rao, 2012). China and India are leading the way in the use of medicinal plants, boasting a combined total of 18,646 species. They are followed by Colombia, South Africa, the United States, and 16 other countries. These percentages range from 7% in Malaysia to 44% in India (Balunas & Kinghorn, 2005; Huang, 2011). In Europe, over 1,300 wild medicinal plants are utilized. Meanwhile, in the United States, it is estimated that approximately 118 of the top 150 prescription drugs are derived from natural sources. This indicates the importance of medicinal plants in modern medicine. It is also worth noting that herbal medicines are a primary source of healthcare for up to 80% of individuals in developing nations, while over 25% of prescribed medicines in developed countries originate from wild plant species (Rafieian-Kopaei, 2011).

Unfortunately, these plant species are on the verge of extinction. Habitat destruction, overharvesting, and the effects of climate change all pose serious threats to these plants (Schipmann et al., 2005). A recent report by the IUCN (International Union for Conservation of Nature) reveals that approximately 50,000 to 80,000 flowering plants have medicinal value, but 15,000 are at risk of extinction. The consequences of this loss are already evident, with 20% of wild plant populations significantly depleted due to human consumption (Bentley, 2010). Therefore, concerted global efforts are essential to address the challenges posed by the depletion of these plant species, promoting a balance between human needs and the conservation of our natural heritage.

DOI: 10.1201/9781032702391-7

7.1.1 Genetic Resources of Medicinal Endangered Plants

Biodiversity conservation itself has multiple challenges including social, political, and scientific. For this IUCN has developed 'Red list of threatened species' under the supervision of both government and civil society organizations. As per the assessment, by the end of December 2023, a total of over 40,000 tree species are listed in the IUCN Red List (https://iucn.org/resources/conservation-tool/iucn-red-list-threatened-species). This also include *Meizotropis pellita, Fritillaria cirrhosa,* and *Dactylorhiza hatagirea* found in the Himalayas. These three plants are included as threatened species (https://www.gktoday.in/himalayan-medicinal-plants-in-iucn-red). Pharmacy states that each plant has certain medicinal importance. Table 7.1 presents the IUCN Red List of Plantae kingdom. This shows a serious concern to conserve these plants. Though there are various factors like anthropogenic activities and environmental factors causing reduction in population sizes. This led to the loss of alleles, causing genetic drift inducing reduction of genetic variation (Waqar et al., 2021). To check this, we need to work on demographic history, successional stages, mating system, pollen dispersal factor and sources, and geographical range. Geographic factors contribute to natural selection, whereas seed and pollen dispersal affect the migration rate of a species (Sork, 2016). There are number of biotechnological and unconventional biotechnological approaches to conserve the genetic resources. Among these, DNA-based and biochemical molecular markers are widely utilized. These markers assist in the identification and classification of organisms, diseases, and specific traits. With reference to isoenzyme markers, molecular markers can efficiently identify clones within the plant population. This is assumed to be an accurate method for homology.

Phytochemicals are the secondary metabolites which is unique to plants and hence act as a marker. These compounds are characterized using advanced analytical technique including high-performance liquid chromatography, gas chromatography, mass spectroscopy, and nuclear magnetic resonance. The presence of these compounds accounts for plant's medicinal properties (Abubakar and Haque, 2020).

Apart from the conventional method, there are various unconventional biotechnological methods that have a central role in genome conservation. Micropropagation is one such *in situ* conservation of plant species for the regeneration of plantlets for pathogen and virus-free production (Khan, Al-Qurainy, and Nadeem, 2012). But for genetic conservation, minimum subculturing should be adopted and choice of explant as axillary but and shoot tip should be preferred. A number of studies support the implication of micropropagation for the isolation of essential oil without affecting somatic variation (Weckx, Inzé, and Maene, 2019).

7.2 CONSERVATION STRATEGIES FOR MEDICINAL PLANTS

Conservation of medicinal plants ensures protection, responsible use, and sustainable use of natural resources (Rajasekharan and Wani, 2020; Nalawade et al., 2003). Implementation of effective regeneration strategies for medicinal endangered plants

TABLE 7.1
Lists of Endangered Medicinal Plants as per the List of IUCN

Name	EX	EW	Subtotal (EX+EW)	CR(PE)	CR(PEW)	Subtotal (EX+EW+ CR(PE)+CR(PEW))	CR	EN	VU	Subtotal (threatened spp.)	LR/cd	NT or LR/nt	LC or LR/lc	DD	Total
Andreaeopsida	0	0	0	0	0	0	0	1	1	2	0	0	0	1	3
Anthocerotopsida	0	0	0	0	0	0	0	2	0	2	0	0	1	0	3
Bryopsida	4	0	4	0	0	4	26	38	32	96	0	22	38	28	188
Bryopsidophyceae	0	0	0	0	0	0	0	0	0	0	0	0	0	1	1
Charophyaceae	0	0	0	0	0	0	0	0	0	0	0	0	11	3	14
Chlorophyceae	0	0	0	0	0	0	0	0	0	0	0	0	0	1	1
Cycadopsida	0	5	5	0	3	8	62	105	71	238	0	41	63	0	347
Florideophyceae	1	0	1	2	0	3	6	0	3	9	0	0	8	60	78
Ginkgoopsida	0	0	0	0	0	0	0	1	0	1	0	0	0	0	1
Gnetopsida	0	0	0	0	0	0	0	1	4	5	0	7	75	10	97
Jungermanniopsida	0	0	0	0	0	0	14	25	21	60	0	11	27	3	101
Liliopsida	15	7	22	83	14	119	888	1,537	1,152	3,577	7	542	4,122	1,022	9,292
Lycopodiopsida	0	0	0	2	0	2	14	12	17	43	0	10	45	13	111
Magnoliopsida	107	32	139	473	31	643	4,583	8,972	8,188	21,743	119	3,256	25,570	4,121	54,948
Marchantiopsida	0	0	0	0	0	0	2	3	2	7	0	1	1	4	13
Pinopsida	0	0	0	1	0	1	30	96	80	206	0	100	300	8	614
Polypodiopsida	2	1	3	8	0	11	71	105	97	273	0	43	290	94	703
Polytrichopsida	0	0	0	0	0	0	0	0	0	0	0	1	0	0	1
Sphagnopsida	0	0	0	0	0	0	6	3	4	13	0	0	3	1	17
Takakiopsida	0	0	0	0	0	0	0	0	1	1	0	0	0	0	1
Ulvophyceae	0	0	0	0	0	0	0	0	0	0	0	0	0	1	1
Total	129	45	174	569	48	791	5,702	10,901	9,673	26,276	126	4,034	30,554	5,371	66,535

IUCN Red List Categories: EX, Extinct; EW, Extinct in the Wild; CR, Critically Endangered (includes CR(PE) and CR(PEW)); EN, Endangered; VU, Vulnerable; LR/cd, Lower Risk/conservation dependent; NT, Near Threatened (includes LR/nt, Lower Risk/near threatened); DD, Data Deficient; LC, Least Concern (includes LR/lc, Lower Risk/least concern) is critical to sustain biodiversity and protect valuable resources. There are two basic approaches to the regeneration of medicinal endangered plants: *in situ* conservation and *ex situ* conservation (Zegeye, 2016).

7.2.1 *In Situ* Conservation

Preserving endangered medicinal plants within their natural habitats is essential, and *in situ* conservation strategies play a crucial role in achieving this. Mainly two strategies are employed to conserve endangered plants in their native environments (Bellon et al., 2017).

7.2.1.1 Natural Conservation

Nature reserves play an important role in *in situ* conservation efforts, serving as sanctuaries that protect our planet's abundant biodiversity. These reserves are strategically designed to protect and conserve the natural habitats of native plant species, especially endangered species. Within these reserves, ecosystems thrive without interference, sustaining the complex web of plant life (Chiarucci, Maccherini, and De Dominicis, 2001). By designating these protected areas, policymakers and conservationists foster an environment where endangered and native plants can prosper, protected from the exposures of human-induced harm, such as habitat destruction and overexploitation (Huang et al., 2002; Rodríguez et al., 2007).

7.2.1.2 Wildlife Nurseries

Wildlife nurseries are crucial locations where various wildlife species breed and nurture their offspring. This is an essential aspect of conservation efforts, as both natural conservations and wildlife nurseries work together to maintain a healthy ecosystem and protect wildlife (Hamilton, 2004; Sharma, Kaur, and Uniyal, 2022). Utilizing wild nurseries has emerged as a promising solution for preserving medicinal plants in their natural habitats. These plants, often found only in specific regions, endangered, and in high demand, can greatly benefit from the safety and protection provided by wild nurseries Laurance et al., 2023).

7.2.2 *Ex Situ* Conservation

It refers to the conservation of an organism outside its natural habitat. In this method, organisms, seeds, or other genetic material are isolated from the source and placed in a controlled environment such as botanical gardens, zoos, seed banks, and other environments (Chaubey, Sharma, and Krishnamurthy, 2015). The main objective is to ensure the survival and genetic diversity of species that may be threatened in their natural habitats due to factors such as habitat degradation, pollution, overexploitation, or other hazards (Avani et al., 2006; Rajpurohit and Jhang, 2015).

7.2.3 *In Vitro* Culture/Micropropagation

The endangered medicinal plants are regenerated by micropropagation techniques (Rout et al., 2006). It is also a valuable technique for plant genetic transformation, the generation of disease-free plants and the production of plant-derived bioactive compounds (Espinosa-Leal et al., 2018). The method of micropropagation undertakes culturing contamination-free transfer of apical or auxiliary buds (explants) on a chemically defined nutrient medium. The cells from explants show totipotency and follow the growth events of differentiation and redifferentiation and arrangement of meristematic centers with the influence of growth regulators. The explant will induce either organogenesis or somatic embryogenesis (Xu & Huang, 2014). The growth events of organ formation or somatic embryo formation may occur by direct or indirect morphogenesis (Rout et al., 2006). The *in vitro* cultured explant of orchid form protocorm-like bodies (PLB), which are regenerated from shoot tip, root tip, or leaf segments through callus formation or directly (Lee et al., 2013). The storage organs like bulbs (onion, garlic), tubers (potato), and rhizomes (lily, turmeric, and ginger) are regenerated when the explants from these storage organs are cultured on nutrient medium (Jo et al., 2008).

The combined use of *in situ* and *ex situ* conservation strategies is a comprehensive approach to promote the regeneration of endangered plants. By integrating both *in situ* and *ex situ* conservation approaches, conservationists can address a broader range of challenges and increase the likelihood of successful regeneration for endangered plants. The key lies in careful coordination, monitoring, and adaptive management strategies to ensure the long-term viability of these plants (Volis & Blecher, 2010). The combined efforts of *in situ* and *ex situ* conservation create a synergistic approach, addressing both immediate threats in natural environments and establishing safety for rare and endangered plants, ultimately contributing to plant species' overall resilience and survival (Subroy et al., 2021).

7.3 BIOTECHNOLOGICAL INTERVENTION FOR CONSERVATION OF THE MEDICINAL ENDANGERED PLANTS

Biotechnological interventions play a crucial role in the conservation of medicinal endangered plants. These approaches leverage advances in biotechnology to propagate, preserve, and restore plant species that are at risk. Here are several biotechnological strategies that can be employed for the conservation of medicinal endangered plants.

7.3.1 Synthetic Seed Technology

The artificially encapsulated plant propagation material is termed as synthetic seeds. It requires encapsulating somatic embryos in a protective coating. There are several advantages of using synthetic seed technology for conservation of medicinally important endangered plants. Synthetic seed manufacturing is one of the large-scale technologies for medicinal plant preservation. Moreover, the technology allows the maintenance of genetic uniformity for many numbers of generations. Moreover, this technology is one of the cheaper methods of raising plants by biotechnological methods. Apart from this, this technology facilitates the rapid multiplication of plants. The largest advantage of using this technology is the direct delivery of plant parts (protected with viable coating) to the field. Moreover, this technology is the short- and medium-term storage method of seeds without losing viability. Moreover, the synthetic seed technology gives an easy method of transporting the medicinally important endangered plants compared to their seedlings.

There are two types of synthetic seeds available for production: desiccated synthetic seed and hydrated synthetic seeds. In desiccated synthetic seeds, the seeds are produced by encapsulating multiple somatic embryos along with the process of desiccation. Desiccation removes the moisture completely from these seeds. The protective coating on somatic embryo is made up of polyoxyethylene (Polyox) as encapsulating material. The polyoxyethylene is non-toxic to embryos and prevents microbial growth. Desiccation process is carried out by decreasing humidity of the environment for one to two weeks or leaving them overnight on the bench. The desiccated seed will be helpful for plant species with somatic embryos tolerant to desiccation stress. The technology of hydrated synthetic seed involves the production of somatic embryos in hydrogel capsules. The hydrated synthetic seed technology is applicable to plant species having desiccation-sensitive somatic embryos. The production of hydrated synthetic seed is carried out by calcium alginate encapsulation (Jang et al., 2020).

7.3.2 Cryopreservation

Cryopreservation has become an important method for conservation of medicinal and endangered plants on a long-term basis since it requires minimum space and maintenance. Moreover, cryopreservation preserves the genetic stability of regenerated plants. It is a method of storing medicinal and endangered plant materials at ultra-low temperatures created by liquid nitrogen. At such temperatures, the metabolic activity of cells ceases, and genetic changes do not occur during storage (Engelmann, 2004). This technique is currently under use in many institutions around the world (Maki et al., 2015).

Vitrification and encapsulation/dehydration are the most popular methods of cryopreservation of medicinal and endangered plant species (Jeyanthi et al., 2011). New cryopreservation methods have been developed by adding new elements to the vitrification-based cryopreservation—encapsulation/vitrification (Wang et al., 2005) and droplet vitrification (Khor et al., 2020). Recent cryopreservation method was published using an aluminum cryo-plate method (Vujović et al., 2015). The aim of cryopreservation protocol is the avoidance of genetic changes in cryopreserved and regenerated medicinal and endangered plants. Fortunately, very small or no differences between cryopreserved and regenerated plants were reported (Maki et al., 2015; Matsumoto et al., 2013). The cryopreservation technology can be applied to either seeds or tubers of medicinal endangered plants.

7.3.3 Seed Banks

During the environmental stress, seeds survive, and germination starts when the environmental stress is removed. Because of the storage of ingredients for survival in embryo, seeds acquire desiccation tolerance mechanism (Wang et al., 2015). Unfortunately, many seed-bearing plants produce desiccation-sensitive seeds, also known as recalcitrant species. Here are some common examples of recalcitrant species—avocado, coconut, and cocoa. Although cryopreservation protocol for recalcitrant seeds is being developed (Walters & Pence, 2020), a recent study identified that seed banking is not a good option for medicinal endangered species (Wyse et al., 2018).

7.3.4 Shoot-Tip Cryopreservation

The root and tuber-based medicinal plant can be preserved by shoot-tip conservation method. This technology is applied for long-time preservation of endangered medicinal plant species such as *Lamprocapnos spectabilis* (L.) Fukuhara (Kulus, 2020), *Rubus humulifolius* C. A. Mey (Edesi et al., 2020), *Pogostemon yatabeanus* (Makino) Press (Lee et al., 2021), *Betula lenta* L. (Rathwell et al., 2016), *Castilleja levisecta* Greenm. (Salama et al., 2018), *Lupinus rivularis* Lindl. (Popova et al., 2021), *Aster altaicus* Willd. var. uchiyamae (Choi et al., 2019), and *Dendrobium moniliforme* (L.) Sw. (Jeyanthi et al., 2011).

7.3.5 Genetic Transformation

The approaches of genetic engineering to bolster the fitness of medicinal plants are not many. In plants, genetic engineering was used to increase the disease resistance feature of the American chestnut trees. A fungus causing blight disease in the North American Chestnut tree came in 1900 from Asia. Research has shown that fungal pathogens secret oxalic acid, which results in becoming the plant susceptible to the disease. Research also showed that oxalate oxidase is an enzyme that breaks down oxalic acid, resulting in killing the pathogen. Researchers cloned a bacteria-secreting oxalate oxidase enzyme in the American chestnut tree, thereby making the tree resistant to the fungal blight disease. This work is one of the breakthrough research projects in the conservation of the medicinal endangered plant through a genetic transformation approach (Jacobs et al., 2023).

7.4 CHARACTERIZATION AND EVALUATION OF MEDICINAL PLANTS AND GERMPLASM PRESERVATION

Characterization and evaluation of medicinal plants, as well as germplasm preservation, are crucial steps in understanding the diversity, quality, and sustainable use of these valuable plant resources. Here are key aspects of the characterization and evaluation process, along with considerations for germplasm preservation.

7.4.1 Morphological and Physiological Traits

Among the ecological factors, the availability of moisture and light determines plant growth and reproduction. The variability in these two ecological factors resulted in the spatial distribution of medicinal endangered plants (Liu et al., 2018).

For example, photoinhibition due to damaging D1 protein formation of PSII is initiated by supersaturated light intensity (Chuyong & Acidri, 2017; Taiz and Zeiger, 2009). The photosynthetic efficiency is also constrained by low light or moisture (Chuyong & Acidri, 2017). To adapt to the climacteric variability of light and soil moisture, plants adjust their morphological and physiological traits (Wu et al., 2017). Hence, documenting the morphological characteristics and physiological traits is extremely crucial for knowing their loss compared to their nearest species. Among the morphological features, documenting the plant height, leaf size, leaf shape, flower color, seed weight, etc. Among the physiological characteristics, knowledge about the growth habits, flowering patterns, and response to environmental stimuli provides knowledge about the adaptability of medicinal plants. The morphological and physiological traits of *Acer catalpifolium* Rehd. (a critically endangered species) and *A. mono* (a widespread same-genus species) were studied by (Zhang et al., 2019). Marked reduction in plant growth, biomass, and physiological characteristics was exhibited in endangered species compared to its widespread same-genus species.

7.4.2 Chemical Composition

It is surprising that plants from which medicines are produced are becoming endangered due to extensive uprooting from the environment. Hence, the chemical characterization of endangered medicinal plants plays a crucial role in species conservation. The loss of bioactive compounds (alkaloids, flavonoids, and terpenoids) contributes to the status of being endangered compared to their widespread same-genus species. Nakurte et al. (2023) studied the chemical composition of wild and micropropagated *Eryngium maritimum* L. plants and found that roots of micropropagated *E. maritimum* had lower total phenolic acid content compared to wild-type plants, but the shoots of micropropagated *E. maritimum* had a similar range of phenolic acid content (Nakurte, Berga, and Mežaka, 2023).

7.4.3 Genetic Diversity

The anthropogenic pressure has altered 33% of the land surface on earth, resulting in habitat loss (Basak et al., 2019). The endangered species are created by habitat fragmentation. Population genetics play a crucial role in assessing the genetic diversity of medicinal and endangered plant species. Nowadays, molecular markers are used to assess the genetic diversity among species. DNA-based marker is the most preferred for conservation and improvement of plant species. There are number of DNA markers having the potential to evaluate genetic diversity (Figure 7.1). Among the molecular markers, the random amplified polymorphic DNA (RAPD), simple sequence repeat (SSR), inter simple sequence repeat (ISSR), amplified fragment length polymorphism (AFLP), and start codon targeted (SCoT) polymorphism markers are widely used in medicinal plant conservation. The usage of different molecular markers for studying the genetic diversity analysis of the medicinal endangered plants is listed in Table 7.2.

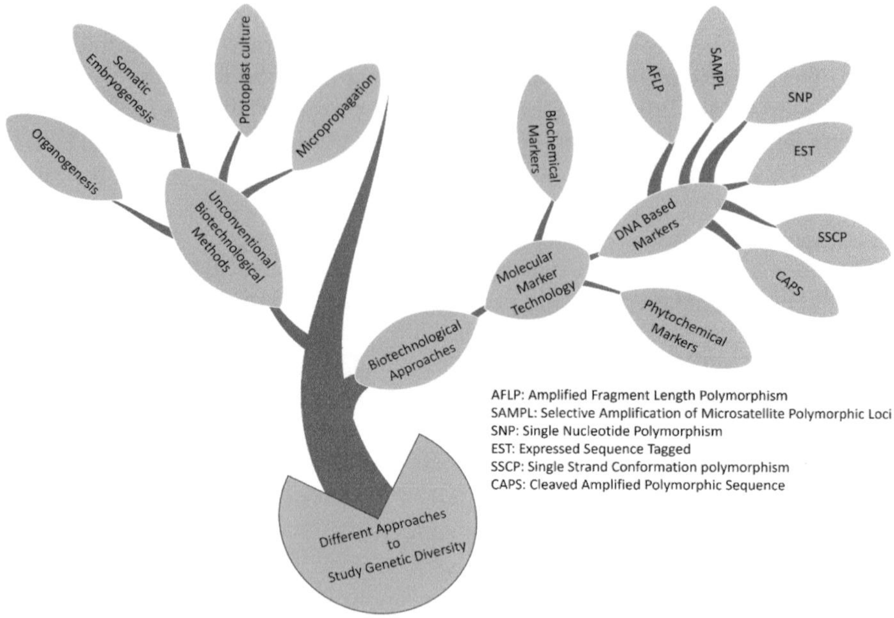

FIGURE 7.1 Different approaches to study genetic variation of medicinal plants and regeneration techniques adopted to preserve genotypes.

TABLE 7.2
Molecular Characterization of Medicinal and Endemic Species

S. No	Medicinal Plant Species	Markers	Reference
1	*Bulbophyllum odoratissimum* (Sm.) Lindl. Ex. Wall.	ISSR	Prasad et al. (2021)
2	*Sideritis euboea Heldr.*	AFLP	Sarrou et al. (2022)
3	*Illicium griffithii*	RAPD, ISSR, DAMD, SCoT	Borah et al. (2021)
4	*Rhododendron pubicostatum*	SSR	Zhang et al. (2020)
5	*Neopicrorhiza scrophulariiflora* Pennel (Hong)	ISSR	Liu et al. (2011)
6	*Ceropegia lawii* Hook	ISSR	Bhamare et al. (2022)
7	Paeonia decomposita Handel-Mazzetti	SSR	Wang (2020)
8	*Coleus forskohlii* Briq.	RAPD, ISSR, AFLP	Tripathi et al. (2013)
9	*Alpinia nigra* (Gaertn.) B.L. Burtt	RAPD, ISSR	Basak et al. (2019)
10	*Curcuma longa* L.	RAPD, ISSR	Basak et al. (2017)

7.5 CONCLUSION

The regeneration of endangered medicinal plants is an important process for the preservation of biodiversity and the protection of traditional knowledge. There are several techniques that can be used for regeneration, including propagation through seed, vegetative parts, or a combination of both. Both *in situ* and *ex situ* conservation strategies and the additional plethora of biotechnological tools are presently being employed to regenerate endangered and medicinal plants.

REFERENCES

Abubakar, A., & Haque, M. (2020). Preparation of medicinal plants: Basic extraction and fractionation procedures for experimental purposes. *Journal of Pharmacy and Bioallied Sciences*, *12*(1), 1. https://doi.org/10.4103/jpbs.JPBS_175_19

Avani, K., Haris, P., & Neeta, S. (2006). Ex situ conservation method for *Clerodendrum inerme*: A medicinal plant of India. *African Journal of Biotechnology*, *5*(5), 415–418.

Balunas, M. J., & Kinghorn, A. D. (2005). Drug discovery from medicinal plants. *Life Sciences*, *78*(5), 431–441. https://doi.org/10.1016/j.lfs.2005.09.012

Basak, S., Chakrabartty, I., Hedaoo, V., Shelke, R. G., & Rangan, L. (2019). Assessment of genetic variation among wild *Alpinia nigra* (Zingiberaceae) population: An approach based on molecular phylogeny. *Molecular Biology Reports*, *46*(1), 177–189. https://doi.org/10.1007/s11033-018-4458-3

Basak, S., Kesari, V., Ramesh, A. M., Rangan, L., Parida, A., & Mitra, S. (2017). Assessment of genetic variation among nineteen turmeric cultivars of Northeast India: Nuclear DNA content and molecular marker approach. *Acta Physiologiae Plantarum*, *39*(2), 45. https://doi.org/10.1007/s11738-016-2341-1

Bellon, M. R., Dulloo, E., Sardos, J., Thormann, I., & Burdon, J. J. (2017). In situ conservation—harnessing natural and human-derived evolutionary forces to ensure future crop adaptation. *Evolutionary Applications*, *10*(10), 965–977. https://doi.org/10.1111/eva.12521

Bentley, R. (2010). *Medicinal Plants*. London: Domville-Fife Press, p. 2346.

Bhamare, M., Thakur, H. A., & Chavan, J. (2022). Efficient plant regeneration and molecular marker-based genetic integrity analysis of Ceropegia lawii Hook: An endemic and endangered herb. *Journal of Applied Biology & Biotechnology*. https://doi.org/10.7324/JABB.2021.100112

Borah, R., Bhattacharjee, A., Rao, S. R., Kumar, V., Sharma, P., Upadhaya, K., & Choudhury, H. (2021). Genetic diversity and population structure assessment using molecular markers and SPAR approach in *Illicium griffithii*, a medicinally important endangered species of Northeast India. *Journal of Genetic Engineering and Biotechnology*, *19*(1), 118. https://doi.org/10.1186/s43141-021-00211-5

Chaubey, O. P., Sharma, A., & Krishnamurthy, G. (2015). Ex-situ conservation of indigenous, threatened and ethno-medicinal diversity of forest species. *International Journal of Bio-Science and Bio-Technology*, *7*(3), 9–22. https://doi.org/10.14257/ijbsbt.2015.7.3.02

Chen, S.-L., Yu, H., Luo, H.-M., Wu, Q., Li, C.-F., & Steinmetz, A. (2016). Conservation and sustainable use of medicinal plants: Problems, progress, and prospects. *Chinese Medicine*, *11*(1), 37. https://doi.org/10.1186/s13020-016-0108-7

Chiarucci, A., Maccherini, S., & De Dominicis, V. (2001). Evaluation and monitoring of the flora in a nature reserve by estimation methods. *Biological Conservation*, *101*(3), 305–314. https://doi.org/10.1016/S0006-3207(01)00073-8

Choi, C. H., Popova, E., Lee, H., Park, S. U., Ku, J., Kang, J. H., & Kim, H.-H. (2019). Cryopreservation of endangered wild species, *Aster altaicus* var. *uchiyamae* Kitam, using droplet-vitrification procedure. *Cryo Letters*, *40*, 113–122.

Chuyong, G. B., & Acidri, T. (2017). Light and moisture levels affect growth and physiological parameters differently in *Faidherbia albida* (Delile) A. Chev. seedlings. *Acta Physiologiae Plantarum*, *39*(5), 117. https://doi.org/10.1007/s11738-017-2410-0

Edesi, J., Tolonen, J., Ruotsalainen, A. L., Aspi, J., & Häggman, H. (2020). Cryopreservation enables long-term conservation of critically endangered species *Rubus humulifolius*. *Biodiversity and Conservation*, *29*(1), 303–314. https://doi.org/10.1007/s10531-019-01883-9

Engelmann, F. (2004). Plant cryopreservation: Progress and prospects. *In Vitro Cellular & Developmental Biology - Plant*, *40*(5), 427–433. https://doi.org/10.1079/IVP2004541

Espinosa-Leal, C. A., Puente-Garza, C. A., & García-Lara, S. (2018). In vitro plant tissue culture: Means for production of biological active compounds. *Planta*, *248*(1), 1–18. https://doi.org/10.1007/s00425-018-2910-1

Hamilton, A. C. (2004). Medicinal plants, conservation and livelihoods. *Biodiversity and Conservation*, *13*(8), 1477–1517. https://doi.org/10.1023/B:BIOC.0000021333.23413.42

Huang, H. (2011). Plant diversity and conservation in China: Planning a strategic bioresource for a sustainable future. *Botanical Journal of the Linnean Society*, *166*(3), 282–300. https://doi.org/10.1111/j.1095-8339.2011.01157.x

Huang, H., Han, X., Kang, L., Raven, P., Jackson, P. W., & Chen, Y. (2002). Conserving native plants in China. *Science*, *297*(5583), 935–936. https://doi.org/10.1126/science.297.5583.935b

Jacobs, D. F., Dumroese, R. K., Brennan, A. N., Campbell, F. T., Conrad, A. O., Delborne, J. A., Fitzsimmons, S., Flores, D., Giardina, C. P., Greenwood, L., Martín, J. A., Merkle, S. A., Nelson, C. D., Newhouse, A. E., Powell, W. A., Romero-Severson, J., Showalter, D. N., Sniezko, R. A., Strauss, S. H., … Woodcock, P. (2023). Reintroduction of at-risk forest tree species using biotechnology depends on regulatory policy, informed by science and with public support. *New Forests*, *54*(4), 587–604. https://doi.org/10.1007/s11056-023-09980-y

Jang, B. K., Cho, J. S., & Lee, C. H. (2020). Synthetic seed technology development and production studies for storage, transport, and industrialization of bracken spores. *Plants*, *9*(9). https://doi.org/10.3390/plants9091079

Jeyanthi, J., James Antony, J., Chan, L., Xavier, R., Marimuthu, S., & Subramaniam, S. (2011). Effect of preculture and PVS2 incubation conditions followed by histological analysis in the cryopreserved PLBs of Dendrobium Bobby Messina orchid. *Australian Journal of Crop Science*, *5*(12), 1557–1564.

Jo, U. A., Murthy, H. N., Hahn, E. J., & Paek, K. Y. (2008). Micropropagation of *Alocasia amazonica* using semisolid and liquid cultures. *In Vitro Cellular & Developmental Biology - Plant*, *44*(1), 26–32. https://doi.org/10.1007/s11627-007-9081-2

Khan, S., Al-Qurainy, F., & Nadeem, M. (2012). Biotechnological approaches for conservation and improvement of rare and endangered plants of Saudi Arabia. *Saudi Journal of Biological Sciences*, *19*(1), 1–11. https://doi.org/10.1016/j.sjbs.2011.11.001

Khor, S. P., Yeow, L. C., Poobathy, R., Zakaria, R., Chew, B. L., & Subramaniam, S. (2020). Droplet-vitrification of Aranda Broga Blue orchid: Role of ascorbic acid on the antioxidant system and genetic fidelity assessments via RAPD and SCoT markers. *Biotechnology Reports*, *26*, e00448. https://doi.org/10.1016/j.btre.2020.e00448

Kulus, D. (2020). Shoot tip cryopreservation of *Lamprocapnos spectabilis* (L.) fukuhara using different approaches and evaluation of stability on the molecular, biochemical, and plant architecture levels. *International Journal of Molecular Sciences*, *21*(11). https://doi.org/10.3390/ijms21113901

Laurance, W. F., Mudhoffir, A. M., Pusparini, W., Meijaard, E., & Engert, J. E. (2023). In Indonesia and beyond nature conservation needs independent science. *Current Biology*, *33*(13), R706–R707. https://doi.org/10.1016/j.cub.2023.04.068

Lee, H.-E., Popova, E., Park, H.-N., Park, S.-U., & Kim, H.-H. (2021). Optimization of a cryopreservation method for the endangered Korean species *Pogostemon yatabeanus* using a systematic approach: The key role of ammonium and growth regulators. *Plants*, *10*(10). https://doi.org/10.3390/plants10102018

Lee, Y.-I., Hsu, S.-T., & Yeung, E. C. (2013). Orchid protocorm-like bodies are somatic embryos. *American Journal of Botany*, *100*(11), 2121–2131. https://doi.org/10.3732/ajb.1300193

Liu, J.-F., Kang, F.-F., Yu, A.-H., Yang, W.-J., Chang, E.-M., & Jiang, Z.-P. (2018). Responses of foliar carbohydrates and nutrient status of two distinctive cypress species to shading and nitrogen addition. *Global Ecology and Conservation*, *16*, e00452. https://doi.org/10.1016/j.gecco.2018.e00452

Liu, X.-L., Qian, Z.-G., Liu, F.-H., Yang, Y.-W., & Pu, C.-X. (2011). Genetic diversity within and among populations of *Neopicrorhiza scrophulariiflora* (Scrophulariaceae) in China, an endangered medicinal plant. *Biochemical Systematics and Ecology*, *39*(4), 297–301. https://doi.org/10.1016/j.bse.2011.03.004

Maki, S., Hirai, Y., Niino, T., & Matsumoto, T. (2015). Assessment of molecular genetic stability between long-term cryopreserved and tissue cultured wasabi (*Wasabia japonica*) plants. *Cryo Letters*, *36*, 318–324.

Matsumoto, T., Akihiro, T., Maki, S., Mochida, K., Kitagawa, M., Tanaka, D., Yamamoto, S., & Niino, T. (2013). Genetic stability assessment of Wasabi plants regenerated from long-term cryopreserved shoot tips using morphological, biochemical and molecular analysis. *Cryo Letters*, *34*, 128–136.

Nakurte, I., Berga, M., & Mežaka, I. (2023). Phytochemical diversity comparison in leaves and roots of wild and micropropagated Latvian sea holly (*Eryngium* mar*itimum* L.). *Molecules*, *28*(9). https://doi.org/10.3390/molecules28093924

Nalawade, S. M., Sagare, A. P., Lee, C.-Y., Kao, C.-L., & Tsay, H.-S. (2003). Studies on tissue culture of Chinese medicinal plant resources in Taiwan and their sustainable utilization. *Botanical Bulletin of Academia Sinica*, *44*(2), 79–98.

Popova, E. V, Shukla, M. R., McIntosh, T., & Saxena, P. K. (2021). In Vitro and cryobiotechnology approaches to safeguard *Lupinus rivularis* douglas ex Lindl., an endangered plant in Canada. *Agronomy*, *11*(1). https://doi.org/10.3390/agronomy11010037

Prasad, G., Seal, T., Mao, A. A., Vijayan, D., & Lokho, A. (2021). Assessment of clonal fidelity and phytomedicinal potential in micropropagated plants of *Bulbophyllum odoratissimum* - An endangered medicinal orchid of Indo Burma megabiodiversity hotspot. *South African Journal of Botany*, *141*, 487–497. https://doi.org/10.1016/j.sajb.2021.05.015

Rafieian-Kopaei, M. (2011). Medicinal plants and the human needs. *Journal of HerbMed Pharmacology*, *1*(1), 1–2.

Rajasekharan, P. E., & Wani, S. H. (2020). Distribution, diversity, conservation and utilization of threatened medicinal plants. In *Conservation and Utilization of Threatened Medicinal Plants* (pp. 3–30). Springer International Publishing. https://doi.org/10.1007/978-3-030-39793-7_1

Rajpurohit, D., & Jhang, T. (2015). In situ and ex situ conservation of plant genetic resources and traditional knowledge. In *Plant Genetic Resources and Traditional Knowledge for Food Security* (pp. 137–162). Springer Singapore. https://doi.org/10.1007/978-981-10-0060-7_8

Rathwell, R., Popova, E., Shukla, M., & Saxena, P. (2016). Development of cryopreservation methods for cherry birch (*Betula lenta* L.), an endangered tree species in Canada. *Canadian Journal of Forest Research*, *46*. https://doi.org/10.1139/cjfr-2016-0166

Rodríguez, J. P., Brotons, L., Bustamante, J., & Seoane, J. (2007). The application of predictive modelling of species distribution to biodiversity conservation. *Diversity and Distributions*, *13*(3), 243–251. https://doi.org/10.1111/j.1472-4642.2007.00356.x

Rout, G. R., Mohapatra, A., & Jain, S. M. (2006). Tissue culture of ornamental pot plant: A critical review on present scenario and future prospects. *Biotechnology Advances*, *24*(6), 531–560. https://doi.org/10.1016/j.biotechadv.2006.05.001

Salama, A., Popova, E., Jones, M. P., Shukla, M. R., Fisk, N. S., & Saxena, P. K. (2018). Cryopreservation of the critically endangered golden paintbrush (*Castilleja levisecta* Greenm.): From nature to cryobank to nature. *In Vitro Cellular & Developmental Biology - Plant*, *54*(1), 69–78. https://doi.org/10.1007/s11627-018-9888-z

Sarrou, E., Doukidou, L., Avramidou, E. V, Martens, S., Angeli, A., Stagiopoulou, R., Fyllas, N. M., Tourvas, N., Abraham, E., Maloupa, E., Nianiou-Obeidat, I., Ganopoulos, I., & Krigas, N. (2022). Chemodiversity is closely linked to genetic and environmental diversity: Insights into the endangered populations of the local endemic plant *Sideritis euboea* Heldr. of Evia Island (Greece). *Journal of Applied Research on Medicinal and Aromatic Plants*, *31*, 100426. https://doi.org/10.1016/j.jarmap.2022.100426

Schipmann, U., Leaman, D. J., Cunningham, A. B., & Walter, S. (2005). Impact of cultivation and collection on the conservation of medicinal plants: Global trends and issues. *Acta Horticulturae*, *676*, 31–44. https://doi.org/10.17660/ActaHortic.2005.676.3

Sharma, R., Kaur, S., & Uniyal, S. Kr. (2022). Population and vulnerability assessment of high value medicinal plants in the Alpine regions of western Himalaya. *Journal of Applied Research on Medicinal and Aromatic Plants*, *26*, 100353. https://doi.org/10.1016/j.jarmap.2021.100353

Simmonds, M. (2005). *Medicinal Plants of the World: Volume 3 Chemical Constituents, Traditional and Modern Medicinal Uses*. Humana Press Incorporated. https://doi.org/10.1016/j.phytochem.2005.11.029

Sork, V. L. (2016). Gene flow and natural selection shape spatial patterns of genes in tree populations: Implications for evolutionary processes and applications. *Evolutionary Applications*, *9*(1), 291–310. https://doi.org/10.1111/eva.12316

Srujana, T. S., Konduri, R. B., & Rao, B. S. S. (2012). Phytochemical investigation and biological activity of leaves extract of plant *Boswellia serrata. The Pharma Innovation Journal*, *1*(5), 22–46.

Subroy, V., Barnes, M., Monks, L., Crawford, A., Coates, D., Pandit, R., Pannell, D., Davis, B., Edgley, M., Gosper, C., & Phillips, B. (2021). *Trade-Offs and Synergies between In Situ and Ex Situ Conservation of Plant Taxa: A Process to Support Practical Decision-Making*. Threatened Species Recovery Hub.

Taiz, L., Eduardo Zeiger, Richard Amasino, Paul Bernasconi, Robert E. Blankenship, Arnold J. Bloom, Ray A. Bressan, Bob B. Buchanan, Daniel J. Cosgrove, and Peter J. Davies. (2010). *Plant Physiology*. Sinauer Associates is an imprint of Oxford University Press.

Tripathi, N., Saini, N., & Tiwari, S. (2013). Morphological and molecular characterization of endangered medicinal plant species Coleus forskohlii collected from central India. *Journal of Crop Science and Biotechnology*, *16*(4), 253–261. https://doi.org/10.1007/s12892-013-0083-9

Volis, S., & Blecher, M. (2010). Quasi in situ: A bridge between ex situ and in situ conservation of plants. *Biodiversity and Conservation*, *19*(9), 2441–2454. https://doi.org/10.1007/s10531-010-9849-2

Vujović, T., Chatelet, P., Ružić, Đ., & Engelmann, F. (2015). Cryopreservation of *Prunus* spp. using aluminium cryo-plates. *Scientia Horticulturae*, *195*, 173–182. https://doi.org/10.1016/j.scienta.2015.09.016

Walters, C., & Pence, V. (2020). The unique role of seed banking and cryobiotechnologies in plant conservation. *Plants, People, Planet*, *3*. https://doi.org/10.1002/ppp3.10121

Wang, Q., Laamanen, J., Uosukainen, M., & Valkonen, J. P. T. (2005). Cryopreservation of in vitro-grown shoot tips of raspberry (*Rubus idaeus* L.) by encapsulation–vitrification and encapsulation–dehydration. *Plant Cell Reports*, *24*(5), 280–288. https://doi.org/10.1007/s00299-005-0936-x

Wang, S.-Q. (2020). Genetic diversity and population structure of the endangered species Paeonia decomposita endemic to China and implications for its conservation. *BMC Plant Biology*, *20*(1), 510. https://doi.org/10.1186/s12870-020-02682-z

Wang, W.-Q., Liu, S.-J., Song, S.-Q., & Møller, I. M. (2015). Proteomics of seed development, desiccation tolerance, germination and vigor. *Plant Physiology and Biochemistry*, *86*, 1–15. https://doi.org/10.1016/j.plaphy.2014.11.003

Waqar, Z., Moraes, R. C. S., Benchimol, M., Morante-Filho, J. C., Mariano-Neto, E., & Gaiotto, F. A. (2021). Gene flow and genetic structure reveal reduced diversity between generations of a tropical tree, *Manilkara multifida* Penn., in Atlantic forest fragments. *Genes*, *12*(12), 2025. https://doi.org/10.3390/genes12122025

Weckx, S., Inzé, D., & Maene, L. (2019). Tissue culture of oil palm: Finding the balance between mass propagation and somaclonal variation. *Frontiers in Plant Science*, *10*. https://doi.org/10.3389/fpls.2019.00722

Wu, J., Li, J., Su, Y., He, Q., Wang, J., Qiu, Q., & Ma, J. (2017). A morphophysiological analysis of the effects of drought and shade on *Catalpa bungei* plantlets. *Acta Physiologiae Plantarum*, *39*(3), 80. https://doi.org/10.1007/s11738-017-2380-2

Wyse, S. V, Dickie, J. B., & Willis, K. J. (2018). Seed banking not an option for many threatened plants. *Nature Plants*, *4*(11), 848–850. https://doi.org/10.1038/s41477-018-0298-3

Xu, L., & Huang, H. (2014). Genetic and epigenetic controls of plant regeneration. In B. Galliot (Ed.), *Current Topics in Developmental Biology* (Vol. 108, pp. 1–33). Academic Press. https://doi.org/10.1016/B978-0-12-391498-9.00009-7

Zegeye, H. (2017). In situ and ex situ conservation: Complementary approaches for maintaining biodiversity. *International Journal of Research in Environmental Studies*, *4*(1), 1–12.

Zhang, X., Qin, H., Xie, W., Ma, Y., & Sun, W. (2020). Comparative population genetic analyses suggest hybrid origin of *Rhododendron pubicostatum*, an endangered plant species with extremely small populations endemic to Yunnan, China. *Plant Diversity*, *42*(4), 312–318. https://doi.org/10.1016/j.pld.2020.06.012

Zhang, Y., Yu, T., Ma, W., Tian, C., Sha, Z., & Li, J. (2019). Morphological and physiological response of *Acer catalpifolium* Rehd. Seedlings to water and light stresses. *Global Ecology and Conservation*, *19*, e00660. https://doi.org/10.1016/j.gecco.2019.e00660

8 Understanding the Role of Epigenetics in Plant Regeneration, Heterosis, and Stress Response

Akash Ajay, Ayushi, and Devinder Kaur

8.1 INTRODUCTION

Like other eukaryotes, even plants undergo heritable changes to their DNA structures without changing the base composition of the DNA sequence, which we know as epigenetic modification (Hamilton, 2011). Epigenetic regulation gives an additional layer of control to eukaryotes and helps them regulate the developmental layout of their bodies, reproduction and also modulate their stress and stimuli response (Skinner, 2011; Han et al., 2019; Gopinathan and Diekwisch, 2022; Ake et al., 2023). Plants display a range of epigenetic modifications that are widely conserved among eukaryotes, while some modifications are distinct and specific to plants (Feng and Jacobsen, 2011). The conservation and divergence of the repertoire of epigenetic changes and histone-modifying enzymes are dependent on the evolutionary history. Plants differ from mammals in the distribution of epigenetic marks on their genome, with repeat regions and transposons overrepresented in epigenetic markings (Feng et al., 2010; Feng and Jacobsen, 2011).

Understanding these epigenetic modifications and their morphological and functional impact is essential to understanding the unique features of plants, such as their regenerative abilities and stress response, as well as some practical applications in generating hybrids that can help in tiding the increasing global nutritional needs. Epigenetics is now understood to modulate the plant response to heat, cold, salt (osmotic), drought, nutrients, and UV stress (Lämke and Bäurle, 2017; Chinnuasamy and Zhu, 2019; Miryeganeh, 2021; Ake et al., 2023). Epigenetic markers such as histone modifications and DNA methylations can generate hybrids that have superior agronomic traits in terms of developmental parameters or stress response and thus be of commercial interest.

The large-scale availability of epigenetic data and the complexity of the processes that drive epigenetic regulation of heterosis, stress, regeneration, and other organismal processes of plants present an opportunity to use modeling approaches to get novel insights. Most of the modeling studies have focused on animal systems with cancer, phenotypic plasticity, senescence, and disease etiology receiving more

 DOI: 10.1201/9781032702391-8

attention; plant epigenetics have begun to catch up with studies focusing on model and non-model plants and epigenetic roles in stress memory and transgenerational inheritance (Khanam Irin et al., 2015; Ruskin and Barat, 2017; Lövkvist and Howard, 2021). This chapter will explain the epigenetics involved in plant regeneration, stress response, and crop management. The generation of vast quantities of epigenetic data in conjunction with other OMICs data has given opportunities for the development of computational approaches. A brief discussion about the same has been covered in this chapter, which we discuss in brief.

8.2 REGENERATION IN PLANTS

Plant regenerative abilities vary from species to species and genotype to genotype (Liu et al., 2023). Plant regeneration could naturally occur in response to injured tissues or may be induced in the lab during tissue culture experiments. Tissue culture regeneration has two types: *de-novo* organogenesis and somatic embryogenesis (Shin et al., 2020; Long et al., 2022; Desai et al., 2022). Somatic embryogenesis involves in the dedifferentiation of plant cells to plant embryonic cells (Yang and Zhang, 2010; Méndez-Hernández et al., 2019; Long et al., 2022). Stress plays a vital role in the initiation of this process *via* hormones like auxins or modifying gene expression. Somatic embryogenesis may be direct (induction of specific somatic cells) or indirect (induction *via* callus, a group of parenchyma cells formed in response to wound) (Lee and Seo, 2018). Direct somatic embryogenesis is faster than indirect ones but gives fewer plantlets than indirect somatic embryogenesis, which is a more straightforward method (Lee and Seo, 2018). *De-novo* organogenesis involves the differentiation of meristematic cells, and no somatic embryo is affected (Lee and Seo, 2018; Shin et al., 2020). The injury-induced and *de-novo* organogenesis regenerative processes differ in their epigenetic regulatory mechanisms and target genes involved (Liu et al., 2023). The gene networks modulating somatic embryogenesis and *de-novo* organogenesis are distinct, and both are regulated epigenetically by similar epigenetic markers of DNA methylation, histone modifications, etc. (Li et al., 2024).

8.3 EPIGENETIC MODIFICATIONS

The propensity and phenotypic effects of epigenetic markers depend on genomic and epigenomic context. There is a vast repertoire of possible epigenetic modifications, and they vary in their functional effect on plant regulation depending on the genomic context (R. Murr, 2010; Feng and Jacobsen, 2011; Miller and Grant, 2012; Lee and Seo, 2018; Hemenway and Gehring 2023). DNA and histones can undergo various sequence-dependent (in terms of nucleotide and the type of residue) chemical modifications that act as the backbone of epigenetic regulatory systems in plants and other species. DNA can get methylated only at the cytosine base position (at the fifth carbon), while the histones can show multiple modifications at multiple positions (mostly at lysine and arginine residues) and can undergo methylation, acetylation, ubiquitylation, phosphorylation, biotinylation, and ADP ribosylation (Figure 8.1) (R. Murr, 2010; Feng and Jacobsen, 2011).

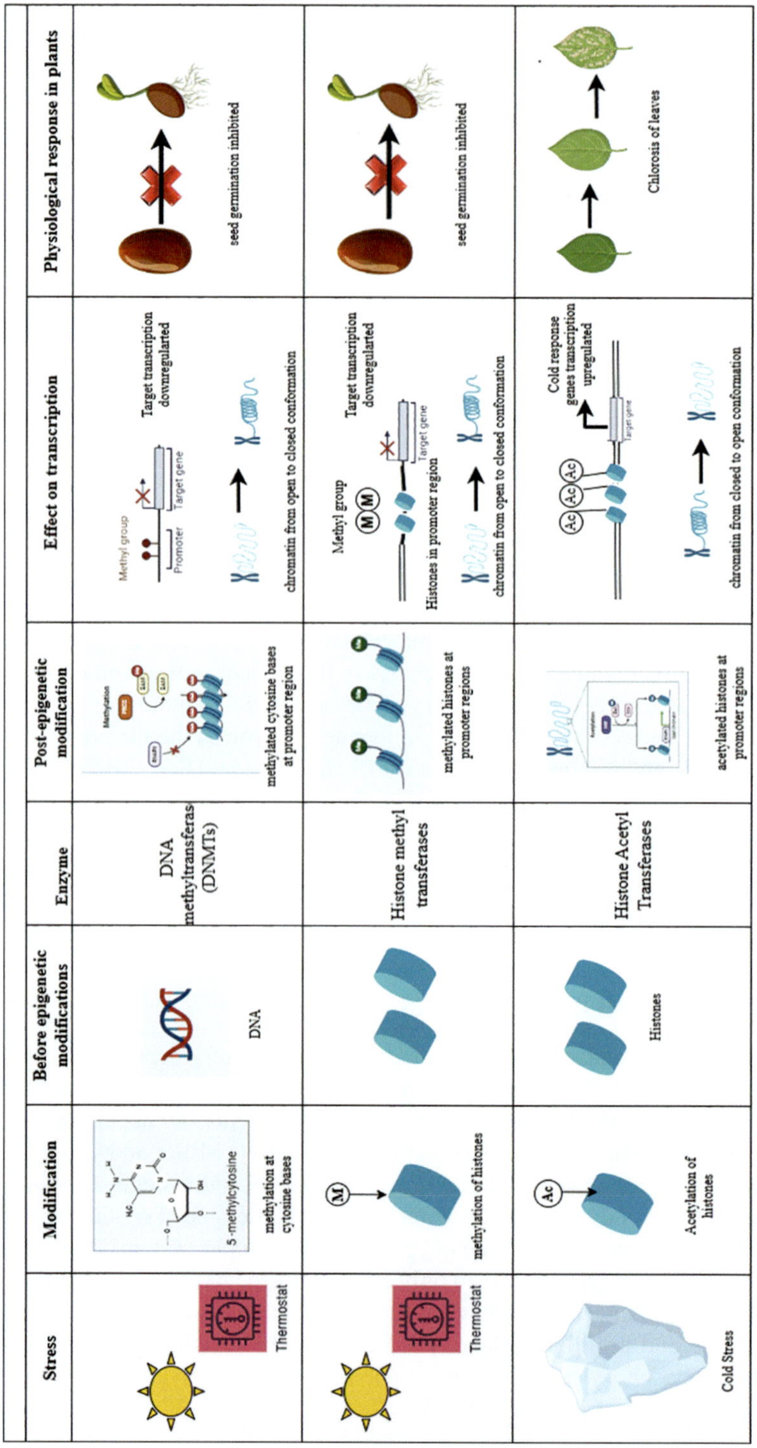

FIGURE 8.1 Plant epigenetic responses to abiotic stress. Schematic representation of the plant response to epigenetic modification in lieu of hot and cold stress.

8.3.1 DNA Methylation in Plants

The methylation enzymes for plants (domains rearranged methyltransferase (DRM2)) are homologous to those in mammals (DNA methyltransferase (Dnmt 3)) (Feng and Jacobsen, 2011). In plants, small RNAs independent of genomic context target the DNA for cytosine methylation in RNA-directed DNA methylation (RdDM) (Matzke et al., 2007; Zhang and Zhu, 2011; Matzke et al., 2015). However, RNA polymerase IV generates ssRNA transcripts, which are mostly complementary to repeat regions and transposons, which act as substrate to RNA-dependent RNA polymerase to generate dsRNA (Feng and Jacobsen, 2011). Polymerase V transcribes intergenic non-coding regions (Wierzbicki et al., 2008). By associating with other proteins, these transcripts may signal DRM2 to target the genomic loci for methylation specifically. By specifically methylating transposon elements (TE) in the genome, the epigenetic pathway acts as an adaptive mechanism for plants to prevent TE proliferation and expression in their genome (Figure 8.1). The maintenance of methylation in plants involves a positive feedback loop between the DNA methylation and the methylation of histone proteins at the methylated genomic regions (Bird, 2021; Cedar and Bergman, 2009; Hashimoto et al., 2010; Feng and Jacobsen, 2011; Rose and Klose, 2014). The enzymes maintaining the methylation in plants are noted to be homologous to their mammalian counterparts (Feng and Jacobsen, 2011). There are differences in the patterns of methylation as well; while plants show methylated cytosines at all sequence contexts, animals show enhanced methylation in CG (CpG dinucleotide) contexts (Law and Jacobsen, 2010; He et al., 2011; Elhamamsy, 2016).

Methylation can be reversed by another epigenetic modification—demethylation, which may be active or passive (Sadakierska-Chudy et al., 2015). Active demethylation is accomplished by DNA glycosylases in plants. In contrast, passive demethylation may be brought upon by reduced expression of enzymes involved in the maintenance of methylated sites (e.g., MET1, which maintains CG methylation levels, is downregulated during oogenesis). It is now believed that the demethylation in plant genomes is a synergistic action of active and passive demethylation (Sadakierska-Chudy et al., 2015; Li et al., 2018; Lee and Seo, 2018; Parrilla-Doblas et al., 2019).

8.3.2 Histone Modifications

A histone is a protein that provides structural support for a chromosome by enabling DNA to wrap around complexes of histone proteins, giving the chromosome a more compact shape. Histones are organized into octamers that are made up of H2A, H2B, H3, and H4. These are linked by H1 histones (Smith, 1991; Schwarz and Hansen, 1994). Histones add to the repertoire of epigenetic modifications called histone posttranslational modifications (HPTM) which include covalent modifications (acetylation, methylation, ubiquitination, phosphorylation, SUMOylation (small ubiquitin like modifications), deamination, adenosine diphosphate (ADP) ribosylation) and non-covalent modifications like proline isomerization (Lee and Seo, 2018; Millán-Zambrano et al., 2022). These posttranslational modifications can lead to the condensation–decondensation of chromatin affecting their accessibility to genomic processes of DNA replication, transcription, and repair and also organize the chromatin into active and inactive regions.

The histones show a greater diversity in the repertoire of epigenetic modifications available (Prakash et al., 2014; Luo et al., 2014; Lee and Seo, 2018) and are known to be responsive to epigenetic changes on DNA strands coiling them and also other epigenetic modifications on nearby histones. Methylated DNA can recruit enzyme complexes that catalyze histone modifications at the genomic loci where the DNA is methylated (Figure 8.1) (Feng and Jacobsen, 2011; Rose and Klose, 2014).

The amino acid composition of the histones affects the possible epigenetic modifications—lysine can undergo methylation at three possible positions while arginine only two, serine and phosphine can undergo phosphorylation reactions, and lysine can undergo acetylation, SUMOylation, and ubiquitination (Smith and Denu, 2009; Li, 2012; Zhao and Garcia, 2015; Alaskhar et al., 2018; Lee and Seo, 2018). The histones, with their repertoire of epigenetic modifications, can alter phenotype by altering the expression levels without any underlying change at the DNA sequence level, giving rise to the concept of "histone code" (Jenuwin and Allis, 2001).

8.3.3 Methylation and Acetylation of Histone Residues

Lysine and arginine work in cellular functions of repressing downstream genes' expression (i.e., silencing) (Li, 2012; Alaskhar et al., 2018; Lee and Seo, 2018). SUMOylation of lysine also works to repress transcription (Lyst and Stancheva, 2007). Histone modifications such as phosphorylation and acetylation activate many cellular responses such as DNA damage response, stress response, and immune response (Figure 8.1) (Chinnusamy and Zhu, 2009; Hegi et al., 2009; Karakaidos et al., 2020; Hannan et al., 2022). Methylation is also known for genomic imprinting in mammals and flowering plants (Li et al., 1993; Finnegan et al., 1998). Imprinted genes are not part of the global demethylation process in which epigenetic tags of the genes are removed; instead, they retain their earlier tags at the gamete stage, and the zygote receives only one working copy of the gene, which may come from either of the parents (Dhar et al, 2021). Methylation has a complex effect on gene expression, while the repressive is common, it is not universal. It is now noted that depending on the context, methylation may have an inhibitory or activating role as the actively transcribed genes are shown to have methylated histones at the cores in enhancer and promoter sites (de-Mendoza et al., 2022).

The individual histone modifications do not operate alone; they work in synchrony (Wang et al., 2016) and show extensive cross-talk with some epigenetic modifications initiating others, for example, HPTMs show their effects on the gene expression by preventing the binding of chromatin proteins and recruiting some other sets of proteins which may have inhibitory or activating effects on the genes downstream, or they may simply alter the local 3D structure of chromatin which may translate to changes in the gene transcription activity for the surrounding genomic regions (Lee and Seo, 2018; Miryeganeh, 2021).

Unlike methylation, acetylation is known for increasing DNA accessibility by creating an open chromatin structure, thereby making acetylation one of the epigenetic markers associated with positive regulation of gene expression (gene activation) (Fu et al., 2007). The addition of a large positive group reduces the electrostatic attraction between the negatively charged DNA and positively charged histone octamers. Acetylation and deacetylation are carried out by histone acetylases and deacetylases, respectively. Samo et al. (2021) show that redox conditions of the plant cells also influence their epigenome. Several factors induce epigenetic changes, among which, one is phytohormones.

8.3.4 The Interplay between Phytohormones and Epigenetics in Regulating Plant Regeneration

Plant regeneration, totipotency, and pluripotency are controlled by phytohormone signaling, and the signaling at the molecular level is facilitated by epigenetic signaling and changes in the chromatin structures (Campos-Rivero et al., 2017; Jiang et al., 2023). In various plants, differences in methylation have been attributed to the changes in cell fate leading to dedifferentiation and formation of callus-like unspecialized structures in which somatic embryos are formed. The auxin-induced methylation enzyme H3K27me3 is observed to be required to facilitate auxin-induced response to initiate callus formation and suppress the leaf identity genes in *Arabidopsis*. Similar experiments with the usage of demethylation blockers have been shown to block the induction of callus sites in peaches as well (Fehér, 2019; Ikeuchi et al., 2019).

Auxin and cytokinin are two important plant growth hormones used in plant regeneration, and these pathways function by interacting with histone modifiers to effect gene expression changes and cellular fates (Li et al., 2013; Mateo-Bonmatí et al., 2019; Wójcikowska et al., 2020). The epigenetic changes so far known to be important in modulating plant regeneration are methylation, acetylation, the occurrence of histone variants, ATP-dependent chromatin remodeling, and the presence of microRNA. Histone acetylation accumulates on genes such as WIND1, ERF113/RAP2.6L, and LBD, which are known to respond to external injury and lead to the formation of calli at the sites of wounding, which is essential for the regeneration of damaged tissue (Liu et al., 2023).

Among all the phases involved in plant regeneration, maximal changes in the global and local epigenome of the cell are noted during the process of callus formation (Lee and Seo, 2018; Liu et al., 2023). The cells in the callus stage are dedifferentiated and thus need to transition to an open chromatin stage to allow diversity in cell fates at later stages of embryo/ tissue regeneration. This predictably leads to a massive loss of DNA methylation at the callus stage. The global acetylation of H3 and H4 histones is enriched at the callus stage. Wounds on plants act as external stimuli and release wound-induced dedifferentiation factors, which are known to initiate callus formation (Lee and Seo, 2018).

Atypical histone variant H3 is noted for its effect on cell fate in *Arabidopsis* and helps in callus formation as it is deficient in sites for methylation (Yan et al., 2020). Chromatin remodelers can affect the transcription activity at very specific genomic locations by altering the number and positions of the nucleosomes at the sites and then increasing or decreasing the chromatin accessibility and transcription activities of cis genes. Chromatin remodeling is one of the mechanisms that increase the rate of callus formation and induction of somatic embryos (Lorch and Kornberg, 2017;

Liu et al., 2023). Micro-RNAs mi156 and mi319 dampen the plant response to the growth hormones auxin and cytokinin by competitively binding to their response genes, thereby reducing the regenerative capabilities of callus with age (Barrera-Rojas et al., 2021). So, the developmental pathways, stress responses, and epigenetics together fine-tune the plant's regenerative capabilities.

8.3.5 Epigenetics in Crop Management—Understanding Heterosis

Heterosis primarily refers to the phenomenon that cross-pollinated plants or hybrid plants having higher fitness levels than individual strains. (Holchdinger and Baldauf, 2018; Wu et al., 2021; Rahman et al., 2021). Heterosis for a trait is measured by the degree of heterosis- the ratio of the hybrid trait values divided by the mean trait value for the parents (East, 1936; Holchdinger and Hoecker, 2007; Holchdinger and Baldauf, 2018). The degree of heterosis differs from trait to trait and from species to species, and the genetic distance and ploidy of the parental strains involved (Holchdinger and Baldauf, 2018; Rehman et al., 2021). Hybrid plants have the lion's share in the global crop market compared to wild-type plants (James, 1997; Holchdinger and Baldauf, 2018). Heterosis remains one of those biotechnological concepts that have found far more prevalence in our day-to-day life and a lesser understanding of the theory of mechanisms driving it. It remains one of those technologies that have a great understanding of the application aspects but a comparatively lesser understanding of the fundamental and theoretical aspects. The hybrid vigor of heterosis also depends upon the genetic distance between the pairs; pairs from genetically dissimilar groups (i.e., heterotic groups) have a greater degree of heterosis than those from the same germplasm groups (Holchdinger and Baldauf, 2018; Rehman et al., 2021). The degree of heterosis and the probability of a successful cross with a fertile hybrid are inversely related. The closer strains are more straightforward to cross, and the hybrids are less likely to be sterile but carry a lesser heterotic effect than the hybrids obtained by crossing distantly related strains or species (Chen, 2013; Rehman et al., 2021).

The molecular explanation of heterosis remains incomplete without incorporating epigenetic aspects. Multiple genomic aspects like trait dominance, overdominance, epistasis, expression patterns, and ploidy were explored to dissect the molecular mechanisms driving heterosis (Shull, 1948; Li et al., 2008; Liang et al., 2015; Hashimoto et al., 2010). These modeling approaches have been found to be insufficient in explaining the progressive heterosis (Birchler et al., 2010), that is polyploidy maximizes the degree of heterosis (Holchdinger and Baldauf, 2018). The limited success of genetic models in giving a complete molecular explanation of heterosis and hybrid vigor has led to researchers trying to incorporate the non-genetic factors contributing to hybrid vigor, among which epigenetics is becoming an active field of research.

8.3.5.1 Epigenetics in Plant Heterosis

Heterozygous plants differ from the parental lines in their epigenetic markings. Apart from parental genetic distance, quantitative genetics, transcriptomics, and proteomics, even epigenetics modifications like DNA methylation, histone modifications, and non-coding RNAs are also hypothesized to contribute to heterosis vigor (Holchdinger and Baldauf, 2018; Ryder et al., 2019; Wu et al., 2021;

Rahman et al., 2021). Homozygosity and heterozygosity in DNA methylation levels modulate inbreeding depression and hybrid vigor (Wu et al., 2021).

Epigenetic mechanisms in the regulation of plant heterosis have been extensively studied for important model and crop plants like *Arabidopsis* (Shen et al., 2012), rice (Chodavarapu et al., 2012), broccoli (Liu et al., 2018a; Li et al., 2018), and maize (Tsafiaris et al., 1999; Chen, 2013). The broccoli plants show increased DNA methylation levels at the hybrids and note it to be proportionate to the germplasmic differences of the parents. Similar findings have also been noted for *Arabidopsis*. These methylations influence the transcription levels of the genes, thereby giving rise to heterosis-induced vigor. One of the ways epigenetic reprogramming contributes to the heterotic effect of hybrids is by affecting the circadian clocks and helping them match with the external environment, thereby enhancing their vigor. Small RNA-directed DNA silencing is noted to be significantly lower in hybrids than the parental strains by many studies on model plants and crops such as *Arabidopsis*, maize, and wheat (Rehman et al., 2021).

Lauss et al. (2018) attempted to quantify the role of epigenetic modifications in heterotic effect by studying the trait variations in genomically similar (isogenic) and epigenetically different parents. The differentially methylated regions between the isogenic parents tended to contribute the most to epigenetic remodeling in hybrids, and locus-specific epigenetic differences were predictive of differences in the trait values.

8.3.6 Epigenetic Modulation of Stress Response

Plants lack motility and a central nervous system to make complex behavioral responses, as is the norm for the animal world to respond to their changing environment. Yet they have continuously thrived since the Ordovician epoch (Miller, 1997) and have helped sustain life on Earth. They evolved complex molecular networks based on epigenetic modifications, enabling them to optimize their stress response and compensate for the lack of central nervous system or mobility (Kovalchuk, 2023). Environmental stress can regulate plant development by regulating the epigenomic states of plant genome thereby making epigenetics a central discussion of the plant stress response. Stress limits resource availability, and with the help of epigenetic remodeling of chromatin, plants can turn off the expression of genes that are not immediately required and boost up the transcription of stress response genes helping plants tide the stress (Figure 8.2).

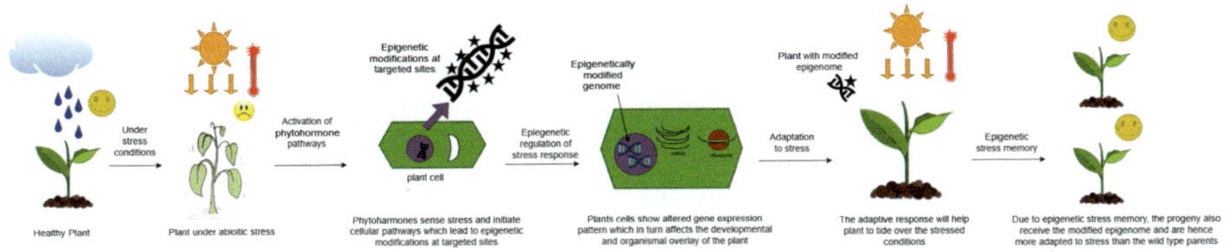

FIGURE 8.2 Plant adaptations to stress and epigenetic stress memory. Schematic representation of epigenetic regulation of stress response in plants and epigenetic stress memory.

8.3.6.1 Epigenetic Stress Memory

Epigenetic stress memory is responsible for a plant's adaptive response to repeated exposure to stress, also known as stress priming (Reul and Chandramohan, 2007; Lämke and Bäurle, 2017; Chinnuasamy and Zhu, 2019). Since plants lack the motility to move away from stress stimuli, molecular mechanisms like stress memory and priming are important survival mechanisms and boost fitness in a changing environment. Both DNA methylation patterns and histone modification patterns contribute to intergenerational memories. Transgenerational memories have helped plants combat osmotic, heavy metal, pathogen, and heat stress (Miryaganeh, 2021). Epigenetic stress memories are not unique to plants and are also noted to occur in mammals (Rakyan et al., 2001; Migicovsky and Kovalchuk, 2011).

8.3.6.2 Epigenetic Modifications Vary with Difference in Stress Resource

Some of the typical stresses that plants undergo are nutrient deficiency, drought, heat, flooding, contamination by pollutants like heavy metals, etc., and epigenetic regulations can help plants mitigate this stress. Heat shock responses of plants typically involve epigenetic modifications like histone posttranscriptional modifications. Heat-tolerant plants are known to differ in their epigenome even under normal conditions, indicating that epigenetic factors are important factors in determining heat sensitivity or resistance. Cytosine methylation and demethylation activities differ between heat-sensitive and resistant variants of *B. napus* (Shen et al., 2017).

Heat stress triggers elongated growth and early flowering in *Arabidopsis*, which has a cooling effect on the plant, and epigenetic changes regulate the expression of these genes (Ake et al., 2023). Epigenetics also plays a role in drought response in *Arabidopsis* via histone acetylation and deacetylation of genes in drought stress pathways such as abscisic acid pathways. A similar study on poplar hybrids showed hypomethylation and hyperacetylation in the epigenome of plants subjected to drought stress (Ake et al., 2019; Miryeganeh, 2021; Khan et al., 2023).

Environmental stress response has been noted to occur via various acetylation–deacetylation mechanisms (Chinnuasamy and Zhu, 2019). Overexpression of histone deacetylases is linked to overexpression of ethylene response factors and pathogenesis response genes (Zhang et al., 2017). In *Arabidopsis*, the deacetylases target the promoters of genes involved in cell proliferation, immune response, and flowering and thereby help in stress tolerance (Tian and Chen, 2001; Long et al., 2006; Chinnuasamy and Zhu, 2019). Cytosine methylation, both symmetric and asymmetric, has been documented in gene repression and tends to be more stable than histone methylation (Talens et al., 2010; Zhou and Robertson, 2016).

Different stresses tend to give different responses in terms of methylation: cold stress in maize and tobacco leads to hypomethylation, while osmotic stress leads to hypermethylation in tobacco. So, the epigenetic changes differ with stress and the plant species involved (Chinnuasamy and Zhu, 2019). RdDM, an important gene silencing mechanism, is elevated in cold stress and reduced in heat stress (Liu et al., 2018b; Guo et al., 2021).

Stress stimulates changes in expression patterns to facilitate judicial usage of limited resources.

Osmotic stress leads to global growth arrest, germination, and seedling establishment for *Arabidopsis* (Figure 8.2) (Sridha and Wu, 2006; Fu et al., 2007). Low-temperature stress arrests the flowering in Arabidopsis by chromatin remodeling the FLOWERING LOCUS C (FLC) gene.

Abiotic stress induces premature senescence in leaves *via* histone deacetylation, thereby reducing photosynthesis activities during the stressed phase. Infection by virus also has an epigenetic response as seen in the case of tobacco mosaic virus (TMV) and tobacco plants, leading to hypomethylation of LRR and TMV resistance genes in the tobacco epigenome, which was transmitted as stress memory to the successive generations, leading to enhanced resistance against the virus for the progeny as well (Boyko et al., 2007, Boyko and Kovalchuk, 2011). Similar stress memories conferring enhanced pathogen resistance are seen in the case of rice plants (Chinnuasamy and Zhu, 2019, Miryaganeh, 2021).

8.3.7 Epigenetics of Species Evolution under Environmental Influence

For a long time, evolutionary biologists and population biologists restricted focusing on the effects of gene sequence or allele on a trait value and how the differences in the prevalence of alleles in a population led to stress response, adaptation, and evolution. Understanding the role of epigenetics at the population level has enhanced our understanding of evolutionary and adaptive processes at the molecular level. It shows the importance of considering the variations of epigenome at the population level. The variants of the same gene with different methylation and histone modification patterns are called epialleles, which vary in expression levels. Epialleles with differential reproductive fitness have an epigenetic influence on downstream evolutionary processes such as natural selection. By modulating plasticity and potentially overcoming genetic constraints, epigenetics can impact acclimatization and is a focal point of contemporary evolutionary research (Vogt, 2017; Ashe et al., 2021; Duarte-Aké et al., 2019). Epigenetic response to environmental stimuli has shaped the ecological divergence of the *Dactylorhiza* species (Paun et al., 2010). The environmental settings can explain the variations in the epigenome; the mangroves in the salt marsh showed a decrease in methylation levels compared to those along the riverside (Lira-Medeiros et al., 2010). The changes in leaf morphology in heterophyllous plants are due to the epigenome, as can be witnessed in the methylation levels of plants producing prickly leaves under herbivore pressure in comparison to wild types (Herera and Bazaga, 2013). Polyploidy a frequent adaptive response in plant lineages (Woodhouse, 2009; Anatskaya and Vinogradov, 2022) needs to be stabilized by overcoming self-incompatibility and meiotic instability that requires synchronization of epigenetic response and genetic responses in hybrids as noted in the case of Arabidopsis allopolyploids (Jiang et al., 2021). By opting for the retention of epigenetic silencing mechanisms even for duplicated genes, plants are able to retain the gene expression patterns for their organismal development and avoid pseudogenization of the duplicated genes (Rapp and Wendel, 2005; Jiang et al., 2021). Another study on Arabidopsis (Latzel et al., 2013) showed that the local adaptation against stress, provided by epigenomes, can help plants wither against climate change and escape possible extinction (Douhovnikoff and Dodd, 2015; Vogt 2017). While studies on speciation epigenetics and epigenomics have primarily focused on animals and protists (Vogt, 2017; Weiner et al., 2023), there is a critical need for more research on the role of epigenetics in plant speciation and adaptation, particularly in non-model species.

8.3.8 Computational Approaches to Study Epigenetic Regulation

In the era of AI/ML, there emerged the field of computational epigenetics, which uses statistical methods and mathematical modeling to explore molecular mechanisms of epigenetics and its association with biological processes. In this section, we will understand the epigenetics as explained by the computational models (Table 8.1, Figure 8.3).

As table 8.1 is already there, before reference section, and is in formatted form, I prefer to edit it there only to not miss any changes. It may be moved here later at your end.

8.3.8.1 Systems-Level Modeling of Epigenetic Models for Heterosis

As noted in the previous sections, the epigenetic changes do not act in isolation and work in synchrony with each other, giving the possibility to explore theoretical models based on epigenetics alone or models with both genetic and epigenetic

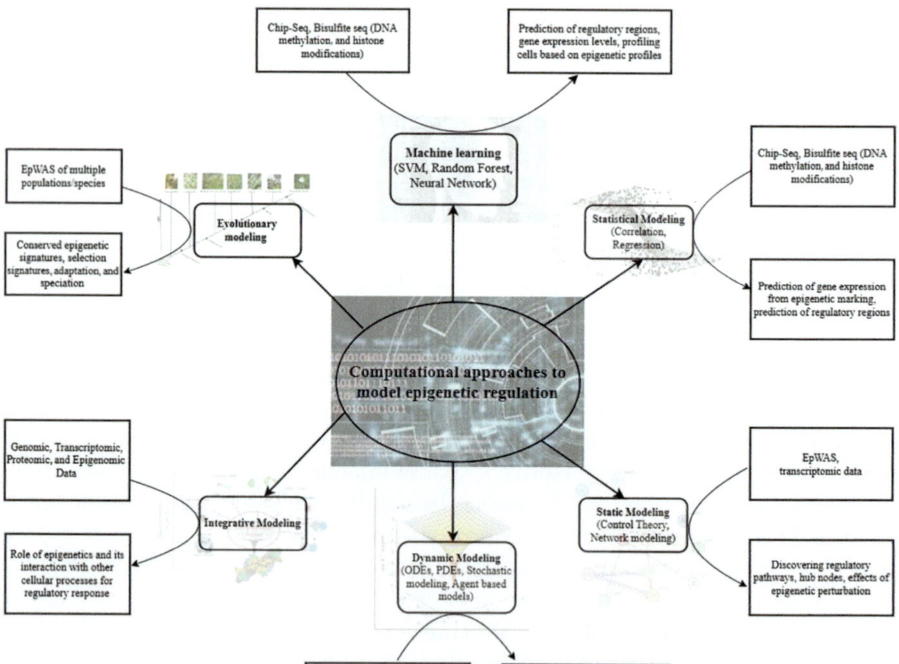

FIGURE 8.3 Computational approaches to model epigenetic regulations. Graphical representation of various computational approaches to model epigenetic regulation in plants.

TABLE 8.1
Computational Approaches to Understand Epigenetic Regulation in Plants

Computational Modeling Technique	Approach	Species	Problem	Insights Gained	Reference
Simulation	Dynamic	*Arabidopsis thaliana*	Cold-induced Verbalization	Verbalization dependent on duration of prior cold	Angel et al. (2011)
Network	Static	*Arabidopsis thaliana*	Tissue-specific gene regulatory networks	Important epigenetic regulators identified	McCoy et al. (2021)
Bayesian	Statistical	*Arabidopsis thaliana*	gene regulatory networks	Incorporation of epigenetic data improves model performance	Zheng et al. (2011)
Comparative epigenomics	Evolutionary	*Arabidopsis thaliana and Zostera marina*	Epimutation clocks	Epimutation clocks recapitulate known phylogenies	Yao et al. (2023)
Integrative modeling	Integrative modeling	*Oryza sativa*	Tissue-specific epigenetic landscapes	Importance of epigenetics in transcriptional regulation	Zhao et al. (2020)
Deep learning	Machine learning	*Oryza sativa and Zea mays*	Prediction of epigenomic sites	Generation of model predicting six types of epigenetic modifications across the genome	Wang et al. (2024)

components to explain the molecular mechanisms behind heterosis. There are differences in methylated cytosine patterns between the hybrids and inbred parental lines, which correspond to gene expression changes at differentially methylated loci. DNA methylation and histone modifications act like signals for each other and generate a positive feedback loop for methylating the loci. Transacting epigenetic modifications like trans chromosomal methylation and trans chromosomal demethylation, which involve alleles mimicking methylation patterns of other alleles when two different epigenomes are hybridizing, make up most of the non-additive methylation differences in heterosis (Greaves et al., 2012; Greaves et al., 2014; Greaves et al., 2016; Ryder et al., 2019).

Ryder et al. (2019) postulate that a series of epigenetic modifications working together as a network would lead to heterosis if it leads to an increased expression of favorable phenotypes and no heterosis if it works otherwise. These epigenetic networks may be one of the factors that drive heterosis at a molecular level. Further experiments coupled with theoretical modeling will help us get a quantitative picture of the role played by epigenetics in driving heterosis and explaining the increased fitness of the hybrid.

8.3.8.2 Computational Modeling of Stress Response

Synchronous occurrence and action of epigenetic markers give rise to emergent behaviors affecting gene transcription and cell regulation, so the systems biology models find widespread applications here. Markov Chain Monte Carlo (MCMC) models and Bayesian models have been used to model the causal relationships between different epigenetic modifications and epigenetic stress memory (Raghavan et al., 2010; Varona et al., 2015; Muyle et al., 2021). Dynamical models using partial differential equations have been used (Roessler et al., 2018; Longoria and Sierra, 2024) to model plant epigenetic responses, especially for model organisms like *Arabidopsis,* for which plenty of data is available. Computational models have allowed us to predict the expression levels of genes and DNA sequences from the information on their epigenetic modifications (Figure 8.3) (Cedar and Bergman, 2009; Ioudinkova et al., 2012). Computational epigenetic models can explain developmental processes like cellular differentiation and highlight the role of epigenetic features in organismal development (Li and Welch, 2023). Computational approaches have given quantitative insights into epigenetic modifications and their relative contributions to the overall network that drives epigenetic silencing and epigenetic stress memories in plants (Antoniou-Kourounioti et al., 2018; Qüesta et al., 2020).

Modeling approaches have been used to study some of the common epigenetic modifications in cells, such as polycomb-mediated gene silencing systems and their roles in epigenetic memory (Lövkvist and Howard, 2021). It is hypothesized that epigenetic memory is maintained with the help of bistable states of histone configuration. AUM model tries to explain the preferential heritability of the histone modifications by noting that histones could be in activating (A) or silencing (M) state, which are autoregulated by self-positive feedback loops and are stabler than the third state—unmodified (U) state leading to passing on the parental epigenetic configuration to daughter cells. These kinds of models have helped understand polycomb switching on FLC genes in *Arabidopsis* (Lokvist et al., 2021). Modeling studies on epigenetics have also explored the 3D chromosomal structure on epigenetic regulations and described the role played by near and distant histone interactions in epigenetic silencing of gene expression (Berry et al., 2017). With the increasing availability of HiC (high-throughput genomic and epigenomic technique to comprehensively detect chromatin interactions in the mammalian nucleus) data and epigenome-wide association study (EWAS) data, more studies are expected to link up epigenetics, 3D genome, gene expression, and evolutionary adaptation of species. Epigenetic regulations, with their heterogeneity of data and a wide variety of biological applications, have emerged as a vibrant field of research.

8.3.9 Conclusion and Future Perspectives

We can see the importance of integrating epigenetic knowledge with genomic, biochemical, and physiological information to understand plant regenerative abilities, stress response, and vigor. Epigenomic compatibility along with genomic compatibility will help in predicting the vigor of the hybrid. Nevertheless, emerging technologies will significantly advance our understanding of the epigenetic regulation of plants by generating new data at a larger scale and at finer resolutions. Precise epigenome information obtained from the emerging single-cell profiling technology (Mulqueen et al., 2019) and predictive tools based on deep learning will allow for a better understanding of the dynamics of epigenome changes during plant development and response to the environment. The single-cell technologies will drive the development of new computational frameworks, generating new insights into the epigenetic process. The third wave of AI brought about by deep learning can be especially helpful in generating new computational frameworks. Incorporating epigenomic information in the systems-level modeling of stress, hybrid vigor, and regenerative response can give better insights than models with genomic/biochemical information alone. Epigenomic models can demonstrate the multi-layer control of plant development and response to stimuli by showing the interaction between hormonal and epigenetic markers, genome with epigenome, apart from showing the synchronization of different epigenetic markers within epigenome.

REFERENCES

Alaskhar Alhamwe, Bilal, Razi Khalaila, Johanna Wolf, Verena Bülow, Hani Harb, Fahd Alhamdan, Charles S. Hii, et al. 2018. "Histone Modifications and Their Role in Epigenetics of Atopy and Allergic Diseases." *Allergy, Asthma and Clinical Immunology*. https://doi.org/10.1186/s13223-018-0259-4.

Anatskaya, Olga V., and Alexander E. Vinogradov. 2022. "Polyploidy as a Fundamental Phenomenon in Evolution, Development, Adaptation and Diseases." *International Journal of Molecular Sciences* 23 (7): 3542.

Angel, Andrew, Jie Song, Caroline Dean, and Martin Howard. 2011. "A Polycomb-Based Switch Underlying Quantitative Epigenetic Memory." *Nature* 476 (7358): 105–108.

Antoniou-Kourounioti, Rea L., Jo Hepworth, Amélie Heckmann, Susan Duncan, Julia Qüesta, Stefanie Rosa, Torbjörn Säll, Svante Holm, Caroline Dean, and Martin Howard. 2018. "Temperature Sensing Is Distributed throughout the Regulatory Network That Controls FLC Epigenetic Silencing in Vernalization." *Cell Systems* 7 (6): 643–55.e9. https://doi.org/10.1016/j.cels.2018.10.011.

Ashe, Alyson, Vincent Colot, and Benjamin P. Oldroyd. 2021. "How does Epigenetics Influence the Course of Evolution?" *Philosophical transactions of the Royal Society B* 376 (1826): 20200111

Barrera-Rojas, Carlos Hernán, Wagner Campos Otoni, and Fabio Tebaldi Silveira Nogueira. 2021. "Shaping the Root System: The Interplay between MiRNA Regulatory Hubs and Phytohormones." *Journal of Experimental Botany*. https://doi.org/10.1093/jxb/erab299.

Berry, Scott, Caroline Dean, and Martin Howard. 2017. "Slow Chromatin Dynamics Allow Polycomb Target Genes to Filter Fluctuations in Transcription Factor Activity." *Cell Systems* 4 (4): 445–457..

Birchler, James A., Hong Yao, Sivanandan Chudalayandi, Daniel Vaiman, and Reiner A. Veitia. 2010. "Heterosis." *The Plant Cell* 22 (7): 2105–2112. https://doi.org/10.1105/tpc.110.076133

Bird, Adrian. 2021. "Methylation Talk between Histones and DNA." *Science* 294 (5549): 2113–5.

Boyko, Alexander, Palak Kathiria, Franz J. Zemp, Youli Yao, Igor Pogribny, and Igor Kovalchuk. 2007. "Transgenerational Changes in the Genome Stability and Methylation in Pathogen-Infected Plants: (Virus-Induced Plant Genome Instability)." *Nucleic Acids Research* 35 (5): 1714–25. https://doi.org/10.1093/nar/gkm029.

Boyko, Alex, and Igor Kovalchuk. 2011. "Genetic and Epigenetic Effects of Plant–Pathogen Interactions: An Evolutionary Perspective." *Molecular Plant* 4 (6): 1014–23. https://doi.org/10.1093/mp/ssr022

Campos-Rivero, Gerardo, Pedro Osorio-Montalvo, Rafael Sánchez-Borges, Rosa Us-Camas, Fátima Duarte-Aké, and Clelia De-la-Peña. 2017. "Plant Hormone Signaling in Flowering: An Epigenetic Point of View." *Journal of Plant Physiology*. https://doi.org/10.1016/j.jplph.2017.03.018.

Cedar, Howard, and Yehudit Bergman. 2009. "Linking DNA Methylation and Histone Modification: Patterns and Paradigms." *Nature Reviews Genetics*. https://doi.org/10.1038/nrg2540.

Chen, Z. Jeffrey. 2013. "Genomic and Epigenetic Insights into the Molecular Bases of Heterosis." *Nature Reviews Genetics*. https://doi.org/10.1038/nrg3503.

Chinnusamy, Viswanathan, and Jian-Kang Zhu. 2009. "Epigenetic Regulation of Stress Responses in Plants." *Current Opinion in Plant Biology* 12 (2): 133–139. https://doi.org/10.1016/j.pbi.2008.12.006.

Chodavarapu, Ramakrishna K., Suhua Feng, Bo Ding, Stacey A. Simon, David Lopez, Yulin Jia, Guo-Liang Wang, Blake C. Meyers, Steven E. Jacobsen, and Matteo Pellegrini. 2012. "Transcriptome and methylome interactions in rice hybrids." *Proceedings of the National Academy of Sciences of the United States of America* 109 (30): 12040–5.

de-Mendoza, Alex, Trung Viet Nguyen, Ethan Ford, Daniel Poppe, Sam Buckberry, Jahnvi Pflueger, Matthew R. Grimmer, et al. 2022. "Large-Scale Manipulation of Promoter DNA Methylation Reveals Context-Specific Transcriptional Responses and Stability." *Genome Biology* 23 (1): 163.

Desai, Parth, Shikha Desai, Rutul Rafaliya, and Ghanshyam Patil. 2022. "Plant Tissue Culture: Somatic Embryogenesis and Organogenesis." In Avinash Chandra Rai, Ajay Kumar, Arpan Modi, and Major Singh (eds.), *Advances in Plant Tissue Culture* (pp. 109–30). Academic Press.

Dhar, Gaurab Aditya, Shagnik Saha, Parama Mitra, and Ronita Nag Chaudhuri. 2021. *DNA Methylation and Regulation of Gene Expression: Guardian of Our Health. Nucleus (India)*. Springer. https://doi.org/10.1007/s13237-021-00367-y.

Douhovnikoff, Vladimir, and Richard S. Dodd. 2015. "Epigenetics: A Potential Mechanism for Clonal Plant Success." *Plant Ecology* 216: 227–233. https://doi.org/10.1007/s11258-014-0430-z

Duarte-Aké, Fátima, Rosa Us-Camas, Víctor J. Cancino-García, and Clelia De-la-Peña. 2019. "Epigenetic Changes and Photosynthetic Plasticity in Response to Environment." *Environmental and Experimental Botany* 159: 108–120.

East, Edward M. 1936. "Heterosis." *Genetics* 21 (4): 375.

Elhamamsy, Amr Rafat. 2016. "DNA Methylation Dynamics in Plants and Mammals: Overview of Regulation and Dysregulation." *Cell Biochemistry and Function*. https://doi.org/10.1002/cbf.3183.

Fehér, Attila. 2019. "Callus, Dedifferentiation, Totipotency, Somatic Embryogenesis: What These Terms Mean in the Era of Molecular Plant Biology?" *Frontiers in Plant Science*. https://doi.org/10.3389/fpls.2019.00536.

Feng, Suhua, and Steven E. Jacobsen. 2011. "Epigenetic Modifications in Plants: An Evolutionary Perspective." *Current Opinion in Plant Biology* 14 (2): 179–186.

Finnegan, E. J., R. K. Genger, K. Kovac, W. J. Peacock, E. S. Dennis. 1998. "DNA Methylation and the Promotion of Flowering by Vernalization." *Proceedings of the National Academy of Sciences of the United States of America* 95 (10): 5824–5829. https://doi.org/10.1073/pnas.95.10.5824.

Fu, Wenqun, Keqiang Wu, and Jun Duan. 2007. "Sequence and Expression Analysis of Histone Deacetylases in Rice." *Biochemical and Biophysical Research Communications* 356 (4): 843–50. https://doi.org/10.1016/j.bbrc.2007.03.010.

Gopinathan, Gokul, and Thomas G.H. Diekwisch. 2022. "Epigenetics and Early Development." *Journal of Developmental Biology*. https://doi.org/10.3390/jdb10020026.

Greaves, Ian K., Michael Groszmann, Elizabeth S. Dennis, and W. James Peacock. 2012. "Trans-Chromosomal Methylation." *Epigenetics* 7 (8): 800–5. https://doi.org/10.4161/epi.20820.

Greaves, Ian K., Michael Groszmann, Aihua Wang, W. James Peacock, and Elizabeth S. Dennis. 2014. "Inheritance of Trans Chromosomal Methylation Patterns from *Arabidopsis* F1 Hybrids." *Proceedings of the National Academy of Sciences of the United States of America* 111 (5): 2017–22.

Greaves, Ian K., Steven R. Eichten, Michael Groszmann, Aihua Wang, Hua Ying, W. James Peacock, and Elizabeth S. Dennis. 2016. "Twenty-Four-Nucleotide siRNAs Produce Heritable Trans-Chromosomal Methylation in F1 Arabidopsis Hybrids." *Proceedings of the National Academy of Sciences of the United States of America* 113 (44): E6895–902. https://doi.org/10.1073/pnas.1613623113

Guo, Wei, Dafang Wang, and Damon Lisch. 2021. "RNA-Directed DNA Methylation Prevents Rapid and Heritable Reversal of Transposon Silencing under Heat Stress in Zea Mays." *PLoS Genetics* 17 (6). https://doi.org/10.1371/journal.pgen.1009326.

Hamilton, James P. 2011. "Epigenetics: Principles and Practice." *Digestive Diseases*, 29: 130–35. https://doi.org/10.1159/000323874.

Han, Qiang, Arthur Bartels, Xi Cheng, Angela Meyer, Yong Qiang Charles An, Tzung Fu Hsieh, and Wenyan Xiao. 2019. "Epigenetics Regulates Reproductive Development in Plants." *Plants*. https://doi.org/10.3390/plants8120564.

Hannan Parker, Adam, Samuel W. Wilkinson, and Jurriaan Ton. 2022. "Epigenetics: A Catalyst of Plant Immunity against Pathogens." *New Phytologist*. https://doi.org/10.1111/nph.17699.

Hashimoto, Hideharu, Paula M. Vertino, and Xiaodong Cheng. 2010. "Molecular Coupling of DNA Methylation and Histone Methylation." *Epigenomics* 2 (5): 657–669. https://doi.org/10.2217/epi.10.44

He, Xin Jian, Taiping Chen, and Jian Kang Zhu. 2011. "Regulation and Function of DNA Methylation in Plants and Animals." *Cell Research*. https://doi.org/10.1038/cr.2011.23.

Hegi, Monika E., Davide Sciuscio, Anastasia Murat, Marc Levivier, and Roger Stupp. 2009. "Epigenetic Deregulation of DNA Repair and Its Potential for Therapy." *Clinical Cancer Research*. https://doi.org/10.1158/1078-0432.CCR-08-1169.

Hemenway, Elizabeth A, and Mary Gehring. 2023. "Epigenetic Regulation during Plant Development and the Capacity for Epigenetic Memory." https://doi.org/10.1146/annurev-arplant-070122.

Herrera, Carlos M., and Pilar Bazaga. 2013. "Epigenetic Correlates of Plant Phenotypic Plasticity: DNA Methylation Differs between Prickly and Nonprickly Leaves in Heterophyllous Ilex Aquifolium (Aquifoliaceae) Trees." *Botanical Journal of the Linnean Society* 171 (3): 441–52. https://doi.org/10.1111/boj.12007.

Hochholdinger, Frank, and Jutta A. Baldauf. 2018. "Heterosis in Plants." *Current Biology* 28 (18): R1089–92. https://doi.org/10.1016/j.cub.2018.06.041

Hochholdinger, Frank, and Nadine Hoecker. 2007. "Towards the Molecular Basis of Heterosis." *Trends in Plant Science*. https://doi.org/10.1016/j.tplants.2007.08.005.

Ikeuchi, Momoko, David S Favero, Yuki Sakamoto, Akira Iwase, Duncan Coleman, Bart Rymen, and Keiko Sugimoto. 2019. "Molecular Mechanisms of Plant Regeneration." *Annual Review of Plant Biology* 70: 377–406. https://doi.org/10.1146/annurev-arplant-050718.

Ioudinkova, Elena S., Ana Barat, Andrey Pichugin, Elena Markova, Ilya Sklyar, Iryna Pirozhkova, Chloe Robin, et al. 2012. "Distinct Distribution of Ectopically Expressed Histone Variants H2A. Bbd and MacroH2A in Open and Closed Chromatin Domains." *PLoS One* 7 (10): e47157.

James, Clive. "Global Status of Transgenic Crops in 1997." *ISAAA Briefs* 5 (1997): 1–31.

Jenuwein, Thomas, and C. David Allis.2001. "Translating the Histone Code." *Science* 293 (5532): 1074–80. https://doi.org/10.1126/science.1063127

Jiang, Kai, Hongwei Guo, and Jixian Zhai. 2023. "Interplay of Phytohormones and Epigenetic Regulation: A Recipe for Plant Development and Plasticity." *Journal of Integrative Plant Biology*. https://doi.org/10.1111/jipb.13384.

Jiang, Xinyu, Qingxin Song, Wenxue Ye, and Z. Jeffrey Chen. 2021. "Concerted Genomic and Epigenomic Changes Accompany Stabilization of Arabidopsis Allopolyploids." *Nature Ecology & Evolution* 5 (10): 1382–1393. https://doi.org/10.1038/s41559-021-01523-y

Karakaidos, Panagiotis, Dimitris Karagiannis, and Theodoros Rampias. 2020. "Resolving DNA Damage: Epigenetic Regulation of DNA Repair." *Molecules* 25 (11) (2020): 2496.

Khan, Aamir Hamid, Ling Min, Yizan Ma, Muhammad Zeeshan, Shuangxia Jin, and Xianlong Zhang. 2023. "High-Temperature Stress in Crops: Male Sterility, Yield Loss and Potential Remedy Approaches." *Plant Biotechnology Journal*. https://doi.org/10.1111/pbi.13946.

Kovalchuk, Igor. 2023. "Role of Epigenetic Factors in Response to Stress and Establishment of Somatic Memory of Stress Exposure in Plants." *Plants* 12 (21): 3667.

Khanam Irin, Afroza, Alpha Tom Kodamullil, Michaela Guendel, and Martin Hofmann-Apitius. 2015. "Computational Modelling Approaches on Epigenetic Factors in Neurodegenerative and Autoimmune Diseases and Their Mechanistic Analysis." *Journal of Immunology Research* 2015 (1): 737168.

Lämke, Jörn, and Isabel Bäurle. 2017. "Epigenetic and Chromatin-Based Mechanisms in Environmental Stress Adaptation and Stress Memory in Plants." *Genome Biology*. https://doi.org/10.1186/s13059 017 1263 6.

Lauss, Kathrin, René Wardenaar, Rurika Oka, Marieke H.A. van Hulten, Victor Guryev, Joost J.B. Keurentjes, Maike Stam, and Frank Johannes. 2018. "Parental DNA Methylation States are Associated with Heterosis in Epigenetic Hybrids." *Plant Physiology* 176 (2): 1627–45. https://doi.org/10.1104/pp.17.01054.

Latzel, V. I. T., Eric Allan, Amanda Bortolini Silveira, Vincent Colot, Markus Fischer, and Oliver Bossdorf. 2013. "Epigenetic Diversity Increases the Productivity and Stability of Plant Populations." *Nature Communications* 4 (1): 2875. https://doi.org/10.1038/ncomms3875

Law, Julie A., and Steven E. Jacobsen. 2010. "Establishing, Maintaining and Modifying DNA Methylation Patterns in Plants and Animals." *Nature Reviews Genetics*. https://doi.org/10.1038/nrg2719.

Lee, Kyounghee, and Pil Joon Seo. 2018. "Dynamic Epigenetic Changes during Plant Regeneration." *Trends in Plant Science*. https://doi.org/10.1016/j.tplants.2017.11.009

Li, C., and J. D. Welch. 2023. "Mathematical Modeling of Epigenetic Gene Regulation during Cell Differentiation." *Nature Biotechnology* 41 (3): 330–331.

Li, En, Caroline Beard, and Rudolf Jaenisch. 1993. "Role for DNA Methylation in Genomic Imprinting." *Nature* 366 (6453): 362–65. https://doi.org/10.1038/366362a0

Li, Hui, Jiye Yuan, Mei Wu, Zhanpin Han, Lihong Li, Hanmin Jiang, Yinglan Jia, et al. 2018. "Transcriptome and DNA Methylome Reveal Insights into Yield Heterosis in the Curds of Broccoli (*Brassica Oleracea* L Var. *Italic*)." *BMC Plant Biology* 18 (1). https://doi.org/10.1186/s12870-018-1384-4.

Li, Jiawen, Qiyan Zhang, Zejia Wang, and Qikun Liu. 2024. "The Roles of Epigenetic Regulators in Plant Regeneration: Exploring Patterns Amidst Complex Conditions." *Plant Physiology* 194 (4): 2022–38. https://doi.org/10.1093/plphys/kiae042

Li, Jing, Xin Nie, Jeanie Li Hui Tan, and Frédéric Berger. 2013. "Integration of Epigenetic and Genetic Controls of Seed Size by Cytokinin in Arabidopsis." *Proceedings of the National Academy of Sciences of the United States of America* 110 (38): 15479–84. https://doi.org/10.1073/pnas.1305175110

Li, Lanzhi, Kaiyang Lu, Zhaoming Chen, Tongmin Mu, Zhongli Hu, and Xinqi Li. 2008. "Dominance, Overdominance and Epistasis Condition the Heterosis in Two Heterotic Rice Hybrids." Genetics 180 (3): 1725–42. https://doi.org/10.1534/genetics.108.091942.

Li, Shisheng. 2012. "Implication of Posttranslational Histone Modifications in Nucleotide Excision Repair." *International Journal of Molecular Sciences* 13 (10): 12461–12486.

Liang, Qingzhi, Lianguang Shang, Yumei Wang, and Jinping Hua. 2015. "Partial Dominance, Overdominance and Epistasis as the Genetic Basis of Heterosis in Upland Cotton (*Gossypium Hirsutum* L.)." *PLoS One* 10 (11). https://doi.org/10.1371/journal.pone.0143548.

Lira-Medeiros, Catarina Fonseca, Christian Parisod, Ricardo Avancini Fernandes, Camila Souza Mata, Monica Aires Cardoso, and Paulo Cavalcanti Gomes Ferreira. 2010. "Epigenetic Variation in Mangrove Plants Occurring in Contrasting Natural Environment." *PLoS One* 5 (4). https://doi.org/10.1371/journal.pone.0010326.

Liu, Chao, Mingbo Wang, Lingli Wang, Qigao Guo, and Guolu Liang. 2018a. "Extensive Genetic and DNA Methylation Variation Contribute to Heterosis in Triploid Loquat Hybrids." *Genome* 61 (6): 437–47. https://doi.org/10.1139/gen-2017-0232

Liu, Gaofeng, Yudong Xia, Tongkun Liu, Shaojun Dai, and Xilin Hou. 2018b. "The DNA Methylome and Association of Differentially Methylated Regions with Differential Gene Expression during Heat Stress in *Brassica Rapa*." *International Journal of Molecular Sciences* 19 (5). https://doi.org/10.3390/ijms19051414.

Liu, Xuemei, Kehui Zhu, and Jun Xiao. 2023. *Recent Advances in Understanding of the Epigenetic Regulation of Plant Regeneration*. ABIOTECH. Springer. https://doi.org/10.1007/s42994-022-00093-2.

Long, Jeff A., Carolyn Ohno, Zachery R. Smith, and Elliot M. Meyerowitz. 2006. "Topless Regulates Apical Embryonic Fate in Arabidopsis." *Science* 312 (5779): 1520–1523.

Long, Yun, Yun Yang, Guangtang Pan, and Yaou Shen. 2022. "New Insights into Tissue Culture Plant-Regeneration Mechanisms." *Frontiers in Plant Science*. https://doi.org/10.3389/fpls.2022.926752.

Lorch, Yahli, and Roger D. Kornberg. 2017. "Chromatin-Remodeling for Transcription." *Quarterly Reviews of Biophysics* 50. https://doi.org/10.1017/S003358351700004X.

Lövkvist, Cecilia, and Martin Howard. 2021. "Using Computational Modelling to Reveal Mechanisms of Epigenetic Polycomb Control." *Biochemical Society Transactions* 49 (1): 71–77. https://doi.org/10.1042/BST20190955.

Luo, Ming, Fu Yu Hung, Songguang Yang, Xuncheng Liu, and Keqiang Wu. 2014. "Histone Lysine Demethylases and Their Functions in Plants." *Plant Molecular Biology Reporter* 32 (2): 558–65. https://doi.org/10.1007/s11105-013-0673-1.

Lyst, M. J., and I. Stancheva. 2007. "A Role for SUMO Modification in Transcriptional Repression and Activation." *Biochemical Society Transactions* 35: 1389–92. https://doi.org/10.1042/BST0351389.

Mateo-Bonmatí, Eduardo, Rubén Casanova-Sáez, and Karin Ljung. 2019. "Epigenetic Regulation of Auxin Homeostasis." *Biomolecules*. https://doi.org/10.3390/biom9100623.

Matzke, Marjori A., Tatsuo Kanno, and Antonius J.M. Matzke. 2015. "RNA-Directed DNA Methylation: The Evolution of a Complex Epigenetic Pathway in Flowering Plants." *Annual Review of Plant Biology* 66 (April): 243–67. https://doi.org/10.1146/annurev-arplant-043014-114633.

Matzke, Marjori, Tatsuo Kanno, Bruno Huettel, Lucia Daxinger, and Antonius J.M. Matzke. 2007. "Targets of RNA-Directed DNA Methylation." *Current Opinion in Plant Biology*. https://doi.org/10.1016/j.pbi.2007.06.007.

McCoy, Rachel M., Russell Julian, Shoban R. V. Kumar, Rajeev Ranjan, Kranthi Varala, and Ying Li. 2021. "A Systems Biology Approach to Identify Essential Epigenetic Regulators for Specific Biological Processes in Plants." *Plants* 10 (2): 364.

Méndez-Hernández, Hugo A., Maharshi Ledezma-Rodríguez, Randy N. Avilez-Montalvo, Yary L. Juárez-Gómez, Analesa Skeete, Johny Avilez-Montalvo, Clelia De-La-Peña, and Víctor M. Loyola-Vargas. 2019. "Signaling Overview of Plant Somatic Embryogenesis." *Frontiers in Plant Science*. https://doi.org/10.3389/fpls.2019.00077.

Migicovsky, Zoë, and Igor Kovalchuk. 2011. "Epigenetic Memory in Mammals." *Frontiers in Genetics*. https://doi.org/10.3389/fgene.2011.00028.

Millán-Zambrano, Gonzalo, Adam Burton, Andrew J. Bannister, and Robert Schneider. 2022. "Histone Post-Translational Modifications — Cause and Consequence of Genome Function." *Nature Reviews Genetics*. https://doi.org/10.1038/s41576-022-00468-7

Miller, Arnold I. 1997. "A New Look at Age and Area: The Geographic and Environmental Expansion of Genera during the Ordovician Radiation." 23. https://doi.org/10.1017/S0094837300019813

Miller, Jaime L., and Patrick A. Grant. 2012. "The role of DNA Methylation and Histone Modifications in Transcriptional Regulation in Humans." In *Epigenetics: Development and Disease*, pp. 289–317. https://doi.org/10.1007/978-94-007-4525-4_13

Miryeganeh, Matin. 2021. "Plants' Epigenetic Mechanisms and Abiotic Stress." *Genes* 12 (8): 1106. https://doi.org/10.3390/genes12081106.

Mulqueen, Ryan M., Brooke A. DeRosa, Casey A. Thornton, Zeynep Sayar, Kristof A. Torkenczy, Andrew J. Fields, Kevin M. Wright, et al. 2019. "Improved Single-Cell ATAC-seq Reveals Chromatin Dynamics of in vitro Corticogenesis." *bioRxiv*: 637256.

Murr, Rabih. 2010. "Interplay between Different Epigenetic Modifications and Mechanisms." *Advances in Genetics* 70: 101–141.

Muyle, Aline, Jeffrey Ross-Ibarra, Danelle K. Seymour, and Brandon S. Gaut. 2021. "Gene Body Methylation Is under Selection in Arabidopsis thaliana." *Genetics* 218 (2): iyab061. https://doi.org/10.1093/genetics/iyab061

Padilla-Longoria, Pablo, and Jesus Sierra. 2024. "On a Stochastic PDE Model for Epigenetic Dynamics," https://arxiv.org/abs/2401.03596.

Parrilla-Doblas, Jara Teresa, Teresa Roldán-Arjona, Rafael R. Ariza, and Dolores Córdoba-Cañero. 2019. "Active DNA Demethylation in Plants." *International Journal of Molecular Sciences*. https://doi.org/10.3390/ijms20194683.

Paun, Ovidiu, Richard M. Bateman, Michael F. Fay, Mikael Hedrén, Laure Civeyrel, and Mark W. Chase. 2010. "Stable Epigenetic Effects Impact Adaptation in Allopolyploid Orchids (Dactylorhiza: Orchidaceae)." *Molecular Biology and Evolution* 27 (11): 2465–73. https://doi.org/10.1093/molbev/msq150.

Prakash, S., R. Singh, and N. Lodhi. 2014. "Histone Demethylases and Control of Gene Expression in Plants." *Cellular and Molecular Biology* 60 (5): 97–105. https://doi.org/10.14715/cmb/2014.60.5.16.

Qüesta, Julia I., Rea L. Antoniou-Kourounioti, Stefanie Rosa, Peijin Li, Susan Duncan, Charles Whittaker, Martin Howard, and Caroline Dean. 2020. "Noncoding SNPs Influence a Distinct Phase of Polycomb Silencing to Destabilize Long-Term Epigenetic Memory at Arabidopsis FLC." *Genes and Development* 34 (5): 446–61. https://doi.org/10.1101/gad.333245.119.

Raghavan, Karthika, Heather J. Ruskin, Dimitri Perrin, Francois Goasmat, and John Burns. 2010. "Computational Micromodel for Epigenetic Mechanisms." *PLoS One* 5 (11): e14031.

Rakyan, Vardhman K., Jost Preis, Hugh D. Morgan, and Emma Whitelaw. 2001. "The Marks, Mechanisms and Memory of Epigenetic States in Mammals." *Biochemical Journal* 356 (1): 1–10.

Rapp, Ryan A., and Jonathan F. Wendel. 2005. "Epigenetics and Plant Evolution." *New Phytologist* 168 (1): 81–91. https://doi.org/10.1111/j.1469-8137.2005.01491.x

Rehman, Attiq ur, Trang Dang, Shanzay Qamar, Amina Ilyas, Reemana Fatema, Madan Kafle, Zawar Hussain, Sara Masood, Shehyar Iqbal, and Khurram Shahzad. 2021. "Revisiting Plant Heterosis—From Field Scale to Molecules." *Genes* 12 (11): 1688. https://doi.org/10.3390/genes12111688.

Reul, Johannes M. H. M., and Yalini Chandramohan. 2007. "Epigenetic Mechanisms in Stress-Related Memory Formation." *Psychoneuroendocrinology* 32: S21–S25.

Roessler, Kyria, Alexandros Bousios, Esteban Meca, and Brandon S. Gaut. 2018. "Modeling Interactions between Transposable Elements and the Plant Epigenetic Response: A Surprising Reliance on Element Retention." *Genome Biology and Evolution* 10 (3): 803–815.

Rose, Nathan R., and Robert J. Klose. 2014. "Understanding the Relationship between DNA Methylation and Histone Lysine Methylation." *Biochim Biophys Acta* 1839 (12): 1362–1372. https://doi.org/10.1016/j.bbagrm.2014.02.007

Ruskin, Heather, and Ana Barat. 2017. "Recent Advances in Computational Epigenetics." *Advances in Genomics and Genetics* 8: 1–12. https://doi.org/10.2147/agg.s115524.

Ryder, Peter, Peter C. McKeown, Antoine Fort, and Charles Spillane. 2019. "Epigenetics and Heterosis in Crop Plants." In Juan Armando Casas-Mollano (ed.), *Epigenetics in Plants of Agronomic Importance: Fundamentals and Applications: Transcriptional Regulation and Chromatin Remodelling in Plants* (pp. 129–147). Springer.

Sadakierska-Chudy, Anna, Richard M. Kostrzewa, and Małgorzata Filip. 2015. "A Comprehensive View of the Epigenetic Landscape Part I: DNA Methylation, Passive and Active DNA Demethylation Pathways and Histone Variants." *Neurotoxicity Research.* https://doi.org/10.1007/s12640-014-9497-5.

Samo, Naseem, Alina Ebert, Joachim Kopka, and Iva Mozgová. 2021. "Plant chromatin, Metabolism and Development–An Intricate Crosstalk." *Current Opinion in Plant Biology* 61: 102002. https://doi.org/10.1016/j.pbi.2021.102002

Schwarz, Patricia M., and Jeffrey C. Hansen. 1994. "Formation and Stability of Higher Order Chromatin Structures. Contributions of the Histone Octamer." *Journal of Biological Chemistry* 269 (23): 16284–16289.

Shen, Huaishun, Hang He, Jigang Li, Wei Chen, Xuncheng Wang, Lan Guo, Zhiyu Peng, et al. 2012. "Genome-Wide Analysis of DNA Methylation and Gene Expression Changes in Two Arabidopsis Ecotypes and Their Reciprocal Hybrids." *The Plant Cell* 24 (3): 875–892.

Shen, Yifei, Shuo Sun, Shuijin Hua, Enhui Shen, Chu-Yu Ye, Daguang Cai, Michael P. Timko, Qian-Hao Zhu, and Longjiang Fan. 2017. "Analysis of Transcriptional and Epigenetic Changes in Hybrid Vigor of Allopolyploid Brassica napus Uncovers Key Roles for Small RNAs." *The Plant Journal* 91 (5): 874–893.

Shin, Jinwoo, Soonhyung Bae, and Pil Joon Seo. 2020. "De Novo Shoot Organogenesis during Plant Regeneration." *Journal of Experimental Botany*. https://doi.org/10.1093/jxb/erz395.

Shull, George Harrison. 1948. "What is" Heterosis"?" *Genetics* 33 (5): 439.

Skinner, Michael K. 2011. "Role of Epigenetics in Developmental Biology and Transgenerational Inheritance." *Birth Defects Research Part C - Embryo Today: Reviews*. https://doi.org/10.1002/bdrc.20199.

Smith, Brian C., and John M. Denu. 2009. "Chemical Mechanisms of Histone Lysine and Arginine Modifications." *Biochimica et Biophysica Acta (BBA)-Gene Regulatory Mechanisms* 1789 (1): 45–57.

Smith, M. Mitchell. 1991. "Histone Structure and Function." *Current Opinion in Cell Biology* 3 (3): 429–37. https://doi.org/10.1016/0955-0674(91)90070-F

Sridha, Sunandini, and Keqiang Wu. 2006. "Identification of AtHD2C as a Novel Regulator of Abscisic Acid Responses in Arabidopsis." *The Plant Journal* 46 (1): 124–33. https://doi.org/10.1111/j.1365-313X.2006.02678.x

Talens, Rudolf P., Dorret I. Boomsma, Elmar W. Tobi, Dennis Kremer, Jw Jukema, Gonneke Willemsen, Hein Putter, P. Eline Slagboom, and Bas Heijmans. 2010. "Variation, Patterns, and Temporal Stability of DNA Methylation: Considerations for Epigenetic Epidemiology." *The FASEB Journal* 24 (9): 3135–3144.

Tian, Lu, and Z. Jeffrey Chen. 2001. "Blocking Histone Deacetylation in Arabidopsis Induces Pleiotropic Effects on Plant Gene Regulation and Development." *Proceedings of the National Academy of Sciences* 98 (1): 200–205.

Tsafiaris, A. S., MKafka, APolidoros, and ETani. Epigenetic Changes in Maize DNA and Heterosis. Genetics and Exploitation of Heterosis in Crops, James G. Coors, Shivaji Pandey. 195–203, American Society of Agronomy, Inc. Crop Science Society of America, Inc., 1999.

Varona, Luis, Sebastián Munilla, Joaquim Casellas, Carlos Moreno, and Juan Altarriba. 2015. "Consequences of Paternally Inherited Effects on the Genetic Evaluation of Maternal Effects." *Genetics Selection Evolution* 47: 1–13.

Vogt, Günter. 2017. "Facilitation of Environmental Adaptation and Evolution by Epigenetic Phenotype Variation: Insights from clonal, Invasive, Polyploid, and Domesticated Animals." *Environmental epigenetics* 3 (1): dvx002.

Wang, Jingjing, Xianwen Meng, Chunhui Yuan, Andrew P. Harrison, and Ming Chen. 2016. "The Roles of Cross-Talk Epigenetic Patterns in *Arabidopsis Thaliana*." *Briefings in Functional Genomics* 15 (4): 278–87. https://doi.org/10.1093/bfgp/elv025.

Wang, Long, Yang Liu, Xiaohan Song, Shiji Wang, Meichun Zhang, Jiayi Lu, Sheng Xu, and Hongyan Wang. 2024. "Ozone Stress-Induced DNA Methylation Variations and Their Transgenerational Inheritance in Foxtail Millet." *Frontiers in Plant Science* 15: 1463584.

Weiner, Agnes K. M., Taylor Sehein, Auden Cote-L'Heureux, Robin S. Sleith, Mattia Greco, Clara Malekshahi, Chase Ryan-Embry, Naomi Ostriker, and Laura A. Katz. 2023. "Single-Cell Transcriptomics Supports Presence of Cryptic Species and Reveals Low Levels of Population Genetic Diversity in Two Testate Amoebae Morphospecies with Large Population Sizes." *Evolution* 77 (11): 2472–2483.

Wierzbicki, Andrzej T., Jeremy R. Haag, and Craig S. Pikaard. 2008. "Noncoding Transcription by RNA Polymerase Pol IVb/Pol V Mediates Transcriptional Silencing of Overlapping and Adjacent Genes." *Cell* 135 (4): 635–48. https://doi.org/10.1016/j.cell.2008.09.035.

Wójcikowska, Barbara, Anna M. Wójcik, and Małgorzata D. Gaj. 2020. "Epigenetic Regulation of Auxin-Induced Somatic Embryogenesis in Plants." *International Journal of Molecular Sciences*. https://doi.org/10.3390/ijms21072307.

Woodhouse, M., D. Burkart-Waco, and L. Comai. 2009. "Polyploidy." *Nature Education* 2 (1): 1.

Wu, Xilin, Yan Liu, Yaowei Zhang, and Ran Gu. 2021. "Advances in Research on the Mechanism of Heterosis in Plants." *Frontiers in Plant Science*. https://doi.org/10.3389/fpls.2021.745726.

Yan, An, Michael Borg, Frédéric Berger, and Zhong Chen. 2020. "The Atypical Histone Variant H3.15 Promotes Callus Formation in Arabidopsis Thaliana." *Development* 147 (11). https://doi.org/10.1242/DEV.184895.

Yang, Xiyan, and Xianlong Zhang. 2010. "Regulation of Somatic Embryogenesis in Higher Plants." *Critical Reviews in Plant Science* 29 (1): 36–57.

Yao, N., Zhilin Zhang, Lei Yu, Rashmi R. Hazarika, C. Yu, Hosung Jang, L. M. Smith, et al. 2023. "An Evolutionary Epigenetic Clock in Plants." *Science* 381 (6665): 1440–1445.

Zhang, Fan, Likai Wang, Bin Qi, Bo Zhao, Eun Esther Ko, Nathaniel D. Riggan, Kevin Chin, and Hong Qiao. 2017. "EIN2 Mediates Direct Regulation of Histone Acetylation in the Ethylene Response." *Proceedings of the National Academy of Sciences* 114 (38): 10274–10279.

Zhang, Huiming, and Jian-Kang Zhu. 2011. "RNA-Directed DNA Methylation." *Current Opinion in Plant Biology* 14 (2): 142–147.

Zhao, Lun, Liang Xie, Qing Zhang, Weizhi Ouyang, Li Deng, Pengpeng Guan, Meng Ma, et al. 2020. "Integrative Analysis of Reference Epigenomes in 20 Rice Varieties." *Nature Communications* 11 (1): 2658.

Zhao, Yingming, and Benjamin A. Garcia. 2015. "Comprehensive Catalog of Currently Documented Histone Modifications." *Cold Spring Harbor Perspectives in Biology* 7 (9). https://doi.org/10.1101/cshperspect.a025064.

Zheng, Jie, Iti Chaturvedi, and Jagath C. Rajapakse. 2011. "Integration of Epigenetic Data in Bayesian Network Modeling of Gene Regulatory Network." In *Pattern Recognition in Bioinformatics: 6th IAPR International Conference, PRIB 2011*, Delft, The Netherlands, November 2–4, 2011. Proceedings 6, pp. 87–96. Springer, Berlin Heidelberg.

Zhou, Dan, and K. D. Robertson. 2016. "Role of DNA Methylation in Genome Stability." In Igor Kovalchuk and Olga Kovalchuk (eds.), *Genome Stability* (pp. 409–424). Academic Press.

9 Applications of Plant Regeneration in Crop Improvement

Acharya Balkrishna, Aakanksha Saini, Priyanka Yadav, Priya Yadav, Brijesh Kumar, and Vedpriya Arya

9.1 INTRODUCTION

Plant regeneration, the process of generating a new plant from a single cell, tissue, or organ, has long been a topic of great interest in crop improvement (Long et al., 2022; Ibáñez et al., 2020). Plant improvement techniques have evolved significantly, progressing from traditional crop domestication to modern scientific plant breeding and DNA-based genotyping and genetic engineering (Jensen, 1983; Perez-de-Castro et al., 2012). Recent advancements in molecular and genomic tools have enabled scientists to pinpoint the genetic changes responsible for the remarkable morphological and other transitions responsible for differentiating crops from their wild progenitors (Smýkal et al., 2018; Perez-de-Castro et al., 2012). In particular, the joint efforts of molecular biologists and plant breeders have contributed to the development of valuable biotechnological methods that can accelerate the improvement of various forage and turf grass species (Kopecký & Studer, 2014; Manzanares et al., 2016).

Extensive studies facilitated by rapid progress in sequencing and resequencing of crop genomes have furthered the understanding of how both the unique population history of cultivated plants and centuries of human selection have impacted their genomes (Smýkal et al., 2018; Tang et al., 2010). The application of conventional pre-genomics scientific breeding methodologies, relying on the utilisation of natural and mutant-induced genetic variation and efficient selection of favourable genetic combinations, has dramatically improved the yield for most major crops over the past century (Prohens, 2011; Spindel & McCouch, 2016). However, the emergence of genomics has provided breeders with a new set of tools and techniques that allow the direct study of the genotype and its relationship to the phenotype, representing a paradigm shift in plant breeding (Prohens, 2011; Dennis et al., 2008).

The integration of biological knowledge and models of gene networks across species, facilitated by a suitable cyber-infrastructure, can generate new layers of knowledge useful for crop improvement. While challenges still exist in the genetic manipulation of recalcitrant monocot species, the impressive progress made in areas

DOI: 10.1201/9781032702391-9

such as the development of transgenic plants, the recovery of intergeneric somatic plants, and the generation of molecular markers for marker-assisted selection demonstrates the importance of plant regeneration techniques in accelerating forage and turf grass improvement (Bevan & Uauy, 2013; Perez-de-Castro et al., 2012).

9.2 MOLECULAR MECHANISMS OF PLANT REGENERATION

Plant regeneration is possible through various mechanisms, including organogenesis, where new organs like shoots or roots form from undifferentiated cells or tissues (Ikeuchi et al., 2019); somatic embryogenesis, in which somatic cells give rise to genetically identical embryos (Dodeman et al., 1997); and protoplast culture, where individual plant cells are isolated and cultivated without cell walls to regenerate whole plants (Davey et al., 2005).

The molecular underpinnings of these plant regeneration processes are complex, involving the interplay of phytohormones, transcription factors, and signalling pathways (Motte et al., 2014). Auxin, for instance, is crucial in initiating shoot and root organogenesis, while cytokinin is essential for shoot formation (Che et al., 2007). The expression of key transcription factors, such as WUSCHEL and LEAFY, is critical for specifying organ identity and maintaining stem cell populations during plant regeneration (Gallois et al., 2004; Schoof et al., 2000).

Recent advancements have explored the molecular mechanisms underlying the regenerative capacity of plant cells. Specifically, the ectopic expression of transcription factors like BBM and WUSCHEL2 has improved the efficiency of regeneration from mature seeds and leaf segments of recalcitrant monocot species like maize (Ikeuchi et al., 2019). These findings suggest that a better understanding of the molecular regulators of plant regeneration can enable the development of more efficient transformation and tissue culture protocols for crop improvement.

9.3 TYPES OF PLANT REGENERATION

9.3.1 *De Novo* Organogenesis

De novo organogenesis, forming new organs such as shoots or roots from undifferentiated cells or tissues, is a crucial process in plant regeneration. This complex developmental pathway often relies on carefully applying exogenous plant growth hormones like cytokinin and auxin to induce the appropriate developmental trajectories (Ikeuchi et al., 2016). Auxin, for instance, is crucial for initiating shoot and root organogenesis, while cytokinin is important for the shoot formation (Su et al., 2011).

The mechanisms underlying *de novo* organogenesis involve the interplay of various transcription factors, such as WUSCHEL and LEAFY, which are critical for specifying organ identity and maintaining stem cells during plant regeneration (Meng et al., 2017; Gallois et al., 2002). Recent advancements have explored the potential of ectopically expressing transcription factors like BBM and WUSCHEL2 to significantly improve regeneration efficiency, even in recalcitrant monocot species like maize (Lowe et al., 2016; Lowe et al., 2018).

The ability to regenerate plants through *de novo* organogenesis has been a valuable tool for crop improvement, particularly in developing improved forage and turf grass varieties (Potrykus, 1990). By optimising the protocols for shoot and root induction, researchers have overcome challenges associated with the recalcitrant nature of these monocot species, leading to the expansion of genetic pool accessible for breeding programs (Bhaskaran and Smith, 1990). The successful application of *de novo* organogenesis in crop improvement often relies on carefully manipulating phytohormones, optimising tissue culture conditions, and identifying key molecular regulators that can enhance regeneration efficiency.

As research continues to advance our knowledge of the molecular mechanisms underlying *de novo* organogenesis, the potential applications of this technology in crop improvement and plant biotechnology are poised for further expansion, enabling the development of novel cultivars with enhanced traits such as stress tolerance, improved nutritional profiles, and increased biomass production.

9.3.2 Somatic Embryogenesis

Somatic embryogenesis is the process by which somatic or non-reproductive cells give rise to genetically identical embryos similar to zygotic embryos formed from fertilisation (Sharma and Millam, 2004). This process can be induced by exposing plant tissues to specific stresses, such as nutrient deprivation or by applying specific plant growth regulators like auxin and cytokinin (Jiménez and Thomas, 2005). The formation of somatic embryos provides a means of rapid and efficient plant propagation, as the regenerated plants are genetically identical clones of the parent plant.

Somatic embryogenesis is a great tool for crop improvement, as it allows for the mass multiplication of elite genotypes and the preservation of valuable germplasm (Ikeuchi et al., 2016). This process has been beneficial in developing improved forage and turf grass varieties, which has helped overcome the challenges associated with the recalcitrant nature of these monocot species (Bai and Qu, 2000). By optimising the protocols for somatic embryo induction and plant regeneration, researchers have been able to develop new cultivars with enhanced traits, such as biotic and abiotic stress tolerance, and improved nutritional quality (Ikeuchi et al., 2013).

The effective application of somatic embryogenesis in crop development often relies on maintaining high regeneration rates and genetic stability during the *in vitro* culture process (Sharma and Millam, 2004). The development of liquid culture systems and the use of alternative gelling agents can significantly improve the productivity and cost-effectiveness of the micropropagation process. Additionally, the use of advanced biotechnological tools, such as bioreactor systems, has further enhanced the large-scale propagation of plants through tissue culture (Steingroewer et al., 2013).

In summary, the application of somatic embryogenesis in crop enhancement has demonstrated significant advancements, facilitated by the development of efficient *in vitro* regeneration protocols and the utilisation of advanced biotechnological tools. These advancements have enabled the large-scale propagation of elite genotypes, the conservation of valuable germplasm, and the expansion of genetic pool available for crop breeding programs, contributing to improved crop cultivars with enhanced agronomic traits.

9.3.3 Protoplast Culture

Protoplast culture is the process of isolating and cultivating individual plant cells without cell walls. It can also be a powerful tool for regenerating whole plants (Bajaj, 1995). This method is beneficial for producing intergeneric somatic hybrids, which combine the desirable traits of two different plant species. By fusing the protoplasts of these genetically distinct plant species, it becomes possible to create novel genetic combinations that may exhibit unique and advantageous characteristics. The propagation of plants from these protoplast-derived somatic hybrids is a critical step, enabling the propagation and evaluation of the newly generated germplasm (Davey et al., 2005).

Protoplast culture provides a versatile platform for the plant breeders and biotechnologists to expand the available genetic pool for crop improvement, facilitating the integration of valuable traits such as stress tolerance, disease resistance, and enhanced nutritional quality (Sihachakr et al., 1997; Bajaj, 1995). This method has been valuable in developing improved forage and turf grass varieties, and it has helped overcome the challenges associated with the recalcitrant nature of these monocot species. By fusing the protoplasts of different grass species, researchers have generated novel hybrid combinations with improved characteristics, such as increased biomass production, enhanced stress tolerance, and superior forage quality.

Furthermore, the production of plants from protoplast-derived somatic hybrids has introduced valuable traits from wild relatives or distant plant species into essential crop species, expanding the genetic pool available for breeding programs. This method has been instrumental in developing disease-resistant or stress-tolerant cultivars, as it incorporates beneficial traits that may not be easily accessible through traditional breeding (Davey et al., 2005).

As research continues to optimise protoplast isolation and regeneration protocols, the applications of this technology in crop development and plant biotechnology are poised for further expansion. By using the potential of protoplast culture, plant breeders and biotechnologists can continue to develop innovative solutions to tackle the challenges faced by modern agriculture for global food security and sustainability.

9.3.4 Protoplast Fusion and Somatic Hybridisation

Protoplast fusion and somatic hybridisation are potent tools for overcoming reproductive barriers and generating novel crop genetic variations. Protoplast fusion involves the fusion of isolated plant cells devoid of their cell walls, allowing for the interchange of genetic material between diverse species or even genera. This technique can benefit the introgression of desirable traits, such as disease resistance or stress tolerance, from wild or related species into cultivated varieties (Grosser & Gmitter, 1990; Davey et al., 2005).

One of the major advantages of protoplast fusion is its potential to bypass the natural reproductive barriers that often prevent the direct hybridisation of distant plant species (Bajaj, 1989). Researchers can create novel genetic combinations that may exhibit unique and advantageous characteristics by fusing the protoplasts of two genetically distinct plant species. This method was valuable in developing improved forage and turf grass varieties, and it has helped overcome the challenges associated with the recalcitrant nature of these monocot species (Vasil, 1994).

Through the fusion of protoplasts, plant breeders and biotechnologists can introduce valuable traits, such as enhanced stress tolerance, disease resistance, or improved nutritional quality, from wild or related species into cultivated crop varieties (Cocking, 2000; Power and Cocking, 1970). The regeneration of whole plants from these protoplast-derived somatic hybrids is a critical step, enabling the propagation and evaluation of the newly generated germplasm (Bajaj, 1994).

The successful application of protoplast fusion and somatic hybridisation in crop improvement often relies on optimising isolation, fusion, and plant regeneration protocols (Davey et al., 2005; Dudits et al., 1995). As research continues to advance in this field, the probable applications of these techniques in plant breeding and biotechnology are poised for further expansion (Vasil, 1994; Bajaj, 1995).

9.4 PRACTICAL APPLICATIONS OF PLANT REGENERATION IN CROP IMPROVEMENT

The development of efficient biotechnological tools for improving cannabis has been hindered by the low regeneration rate of transgenic plants and the low production of secondary metabolites in hairy root and cell suspension cultures. This underscores the need for further research to optimise regeneration protocols for specific plant species.

Despite these species-specific challenges, the successful development of efficient plant regeneration systems, combined with the ability to produce transgenic plants and recover intergeneric hybrids, has been helpful in accelerating the improvement of various crop species, including forage and turf grasses, as well as medicinal plants like cannabis. These advancements in plant biotechnology allow for prospects like introducing valuable traits, preserving endangered species, and sustainably producing high-value plant based products.

The application of plant regeneration techniques, such as organogenesis, somatic embryogenesis, and protoplast culture, has played a critical role in advancing crop improvement and plant biotechnology. When combined with the expanding knowledge of plant developmental biology and molecular genetics, these powerful tools can unlock new frontiers in sustainable agriculture, environmental conservation, and the production of high-value plant-derived compounds.

Plant regeneration methods have numerous applications in crop improvement, including developing genetically modified crops, generating novel germplasm through somatic hybridisation, conserving and propagating rare or endangered plant species, and developing new and improved forage and turf grass varieties. These methods are instrumental in rapidly multiplying new crop genotypes and eliminating pathogens through micropropagation, which is crucial for maintaining the genetic and phenotypic integrity of elite crop lines.

9.4.1 Micropropagation

The successful application of plant regeneration in crop improvement relies on developing and optimising efficient in vitro regeneration protocols. Micropropagation, the rapid multiplication of plants through tissue culture techniques, has emerged as a great tool for the large-scale propagation of a variety of plant species, like cash crops, ornamentals, and medicinal plants (Nirmal et al., 2023; Reuther, 1990). Micropropagation offers many advantages over conventional propagation methods, such as the ability to produce many genetically identical plants in a short period, the potential for year-round production, and the opportunity to maintain and propagate disease-free, elite plant material (Murthy et al., 2023; Nirmal et al., 2023; Reuther, 1990).

One of the key challenges in micropropagation is the development of cost-effective and scalable production systems. Traditional micropropagation methods that depend on solid culture media and manual handling can be labour-intensive and expensive. To address these limitations, advancements have been made in using liquid culture systems and bioreactor technologies (Nirmal et al., 2023; Murthy et al., 2023). Liquid culture systems with the suspension of plant tissues in a nutrient-rich medium can enhance the reproducibility and efficiency of the micropropagation process (Nirmal et al., 2023). Also, bioreactors provide a controlled environment for large-scale plant propagation, offering benefits such as improved gas exchange, precise control of environmental factors, and the potential for automation (Nirmal et al., 2023).

Integrating these advanced propagation techniques with other complementary technologies, such as hydroponics and aeroponics, can enhance plant regeneration systems' efficiency and cost-effectiveness (Cardoso et. al., 2018) (Murthy et al., 2023). Combining micropropagation with these innovative approaches makes it possible to streamline the production process, reduce labour and material costs, and improve the quality of the regenerated plants.

The application of plant regeneration in crop enhancement is not limited to the propagation of elite genotypes. It is also crucial for developing better and improved crop varieties through genetic engineering and hybridisation. The ability to regenerate plants from genetically modified or hybridised tissues is important for evaluating and selecting desirable traits, and the subsequent multiplication and deployment of the improved plant materials.

In conclusion, applying plant regeneration methods, such as organogenesis, somatic embryogenesis, and protoplast culture, holds immense potential for enhancing crop improvement. By using these powerful tools, plant breeders and biotechnologists can expand the available genetic diversity, introduce valuable traits, and propagate improved plant materials on a large scale. The continued development and optimisation of efficient, cost-effective, and scalable plant regeneration systems will be instrumental in tackling the hurdles faced by modern agriculture, ultimately leading to global food security and the sustainable production of high-quality crops.

9.4.2 Bioreactor-Mediated Regeneration

Researchers have increasingly adopted bioreactor systems to facilitate the large-scale propagation of plants through tissue culture, as these systems offer several advantages over traditional semisolid culture methods (Escalona et al., 1999; Georgiev et al., 2014). Bioreactors can provide a more controlled and optimised environment for plant growth and development, allowing for the efficient delivery of essential nutrients, gases, and other factors to the growing propagules. As described in the literature, using liquid culture systems in bioreactors has emerged as a particularly efficient approach for sustainable micropropagation (Georgiev et al., 2014).

Liquid culture systems in bioreactors can overcome the limitations of agar-based semisolid media, which can restrict gas exchange, nutrient absorption, and the overall growth and development of in vitro-generated plantlets. Using plantlets and systems in bioreactors has been shown to enhance the multiplication rate, improve the quality of the regenerated plantlets, and increase their survival when transferred to ex vitro conditions (Georgiev et al., 2014).

9.5 CHALLENGES AND FUTURE PROSPECTS

Despite the numerous successes and widespread applications of plant regeneration in crop improvement, several challenges must be addressed. One of the key challenges is the optimisation of cost-effective and efficient in vitro propagation systems for commercially important crops (Georgiev et al., 2014; Hussain et al., 2012). This is crucial for enabling large-scale commercial production and dissemination of desirable plant varieties, ensuring that farmers and growers can widely access the benefits of these improved traits (Altpeter et al., 2016).

Another significant challenge is to enhance the success rate of genetic transformation and the subsequent regeneration of transgenic plants, particularly in recalcitrant monocot species such as forage and turf grasses (Sharma et al., 2020; Teixeira da Silva et al., 2016). Overcoming the recalcitrance of these species to genetic manipulation is a longstanding hurdle, and continued research is necessary to develop better protocols for their transformation and regeneration (Wójcik et al., 2020). Improving the efficiency of these processes will unlock the potential for introducing valuable traits, such as increased stress tolerance and disease resistance, into these economically important grass species.

Ongoing research efforts focus on gaining a better knowledge of the molecular mechanisms underlying plant regeneration, enabling more precise manipulation and control of the process. This includes investigating the regulatory pathways and key genes involved in determining plant cell fate and the transition from undifferentiated to differentiated states (Ikeuchi et al., 2019). By elucidating these critical genetic and molecular components, researchers can develop more targeted strategies to enhance and optimise plant regeneration (Kareem et al., 2016). Additionally, integrating advanced computational approaches, such as machine learning and artificial intelligence, may aid in developing predictive models to refine further and streamline plant regeneration protocols.

Despite these remaining challenges, the prospects of plant regeneration in crop improvement are highly promising. As researchers continue to explore the boundaries of this technology, we can expect even more transformative applications in the years to come, leading to innovative and sustainable agricultural solutions to meet future demand for food, fuel, and other plant-derived products (Sattiraju et al., 2018; Sharma et al., 2002).

9.6 CONCLUSION

The techniques of plant regeneration have become indispensable in the field of crop improvement. The ability to regenerate whole plants from undifferentiated cells or tissues has enabled a wide range of advancements, including the development of genetically modified crops, the generation of novel germplasm through somatic hybridisation, the rapid multiplication of desirable genotypes, the elimination of pathogens, and the conservation of endangered medicinal and aromatic plants. These powerful biotechnological tools have revolutionised how we approach crop improvement, allowing us to unlock new avenues for enhancing agricultural productivity and sustainability. However, it is crucial to note that the efficiency of plant regeneration can vary significantly among plant species. This variability necessitates optimising species-specific protocols to harness these techniques' potential fully. Ongoing research efforts focus on gaining a better knowledge of the underlying molecular mechanisms, enabling more precise manipulation and control of the plant regeneration process. By overcoming these challenges and continuing to refine the application of plant regeneration, we can unlock new frontiers in crop improvement and secure a more promising future for global food production.

REFERENCES

Altpeter, F., Springer, et al. 2016. "Advancing crop transformation in the era of genome editing." *The Plant Cell 28*(7): 1510–1520.

Bai, Y., & Qu, R. 2000. "An evaluation of callus induction and plant regeneration in twenty-five turf-type tall fescue (*Festuca arundinacea* Schreb.) cultivars." *Grass and Forage Science 55*(4): 326–330.

Bajaj, Y. P. S. 1989. "Cryopreservation of plant protoplasts." In Bajaj, Y.P.S. (eds) *Plant Protoplasts and Genetic Engineering I*. Springer, Berlin, Heidelberg: 97–106. https://doi.org/10.1007/978-3-642-73614-8_6

Bajaj, Y. P. S. 1994. "Somatic hybridisation—a rich source of genetic variability." In Bajaj, Y.P.S. (ed) *Somatic Hybridisation in Crop Improvement I*. Biotechnology in Agriculture and Forestry, vol 27. Springer, Berlin, Heidelberg: 3–32. https://doi.org/10.1007/978-3-642-57945-5_1

Bajaj, Y. P. S. 1995. "Cryopreservation of plant cell, tissue, and organ culture for the conservation of germplasm and biodiversity." In Bajaj, Y.P.S. (ed) *Cryopreservation of Plant Germplasm I*. Biotechnology in Agriculture and Forestry, vol 32. Springer, Berlin, Heidelberg: 3–28. https://doi.org/10.1007/978-3-662-03096-7_1

Bevan, M. W., & Uauy, C. 2013. "Genomics reveals new landscapes for crop improvement." *Genome Biology*, 14(6): 1–11.

Bhaskaran, S., & Smith, R. H. 1990. "Regeneration in cereal tissue culture: a review." *Crop science 30*(6): 1328–1337.

Cardoso, J. C., Gerald, L. T. S., & Silva, J. A. T. D. 2018. "Micropropagation in the twenty-first century." In Loyola-Vargas, V., Ochoa-Alejo, N. (eds) *Plant Cell Culture Protocols* Methods in Molecular Biology, vol 1815. Humana Press, New York, NY: 17–46. https://doi.org/10.1007/978-1-4939-8594-4_2

Che, P., Lall, S., & Howell, S. H. 2007. "Developmental steps in acquiring competence for shoot development in Arabidopsis tissue culture." *Planta 226*: 1183–1194.

Cocking, E. C. 2000. "Plant protoplasts." *In Vitro Cellular & Developmental Biology. Plant 36*(2): 77–82.

Davey, M. R., Anthony, P., Power, J. B., & Lowe, K. C. 2005. "Plant protoplasts: status and biotechnological perspectives." *Biotechnology Advances 23*(2): 131–171.

Dennis, E. S., Ellis, J. G., Green, A., Llewellyn, D., Morell, M K., Tabe, L., & Peacock, W. J. 2008. "Genetic contributions to agricultural sustainability." *Philosophical Transactions of the Royal Society B: Biological Sciences* 363(1491): 591–609.

Dodeman, V. L., Ducreux, G., & Kreis, M. 1997. "Zygotic embryogenesis versus somatic embryogenesis." *Journal of Experimental Botany 48*(8): 1493–1509.

Dudits, D., Györgyey, J., Bögre, L., & Bakó, L. 1995. "Molecular biology of somatic embryogenesis." In Thorpe, T.A. (eds) *In vitro Embryogenesis in Plants*. Current Plant Science and Biotechnology in Agriculture, vol 20. Springer, Dordrecht. https://doi.org/10.1007/978-94-011-0485-2_.

Escalona, M., Lorenzo, J. C., González, B., Daquinta, M., González, J. L., Desjardins, Y., & Borroto, C. G. 1999. "Pineapple (*Ananas comosus* L. Merr) micropropagation in temporary immersion systems." *Plant Cell Reports 18*: 743–748.

Gallois, J. L., Nora, F. R., Mizukami, Y., & Sablowski, R. 2004. "WUSCHEL induces shoot stem cell activity and developmental plasticity in the root meristem." *Genes & Development 18*(4): 375–380.

Gallois, J. L., Woodward, C., Reddy, G. V., & Sablowski, R. 2002. "Combined SHOOT MERISTEMLESS and WUSCHEL trigger ectopic organogenesis in Arabidopsis." *Development* 129: 3207–3217.

Georgiev, V., Schumann, A., Pavlov, A., & Bley, T. 2014. "Temporary immersion systems in plant biotechnology." *Engineering in Life Sciences 14*(6): 607–621.

Grosser, J. W., & Gmitter Jr, F. G. 1990. "Protoplast fusion and citrus improvement." *Plant Breeding Reviews 8*: 339–374.

Hussain, A., Qarshi, I. A., Nazir, H., & Ullah, I. 2012. "Plant tissue culture: current status and opportunities." *Recent Advances in Plant In Vitro Culture 6*(10): 1–28.

Ibáñez, S., Carneros, E., Testillano, P S., & Pérez-Pérez, J M. 2020. "Advances in plant regeneration: shake, rattle and roll." *Plants* 9(7): 897–897.

Ikeuchi, M., Favero, D S., Sakamoto, Y., Iwase, A., Coleman, D., Rymen, B., & Sugimoto, K. 2019. "Molecular mechanisms of plant regeneration." *Annual Reviews* 70(1): 377–406.

Ikeuchi, M., Ogawa, Y., Iwase, A., & Sugimoto, K. 2016. "Plant regeneration: cellular origins and molecular mechanisms." *Development 143*(9): 1442–1451.

Ikeuchi, M., Sugimoto, K., & Iwase, A. 2013. "Plant callus: mechanisms of induction and repression." *The Plant Cell 25*(9): 3159–3173.

Jensen, N F. 1983. "Crop breeding as a design science." In Wood D.R. (ed.) *Crop Breeding*: American Society for Agronomy & Crop Society of America, Madison, WI, 21–29. https://doi.org/10.2135/1983.cropbreeding.c2

Jiménez, V. M., & Thomas, C. 2005. "Participation of plant hormones in determination and progression of somatic embryogenesis." In Mujib, A., Šamaj, J. (eds) *Somatic Embryogenesis*. Plant Cell Monographs, vol 2. Springer, Berlin, Heidelberg: 103–118. https://doi.org/10.1007/7089_034

Kareem, A., Radhakrishnan, D., Sondhi, Y., Aiyaz, M., Roy, M. V., Sugimoto, K., & Prasad, K. 2016. "De novo assembly of plant body plan: a step ahead of Deadpool." *Regeneration 3*(4): 182–197.

Kopecký, D., & Studer, B. 2014. "Emerging technologies advancing forage and turf grass genomics." *Biotechnology Advances 32*(1): 190–199.

Long, Y., Yang, Y., Pan, G., & Shen, Y. 2022. "New insights into tissue culture plant-regeneration mechanisms." *Frontiers in Plant Science 13*: 926752.

Lowe, K., et al. 2018. "Rapid genotype "independent" Zea mays L. (maize) transformation via direct somatic embryogenesis." *In Vitro Cellular & Developmental Biology-Plant 54*: 240–252.

Lowe, K., et al. 2016. "Morphogenic regulators Baby boom and Wuschel improve monocot transformation." *The Plant Cell 28*(9): 1998–2015.

Manzanares, C., Yates, S., Ruckle, M E., Nay, M M., & Studer, B. 2016. "TILLING in forage grasses for gene discovery and breeding improvement." *New Biotechnology 33*(5): 594–603.

Meng, W. J., et al. 2017. "Type-B ARABIDOPSIS RESPONSE REGULATORs specify the shoot stem cell niche by dual regulation of WUSCHEL." *The Plant Cell 29*(6): 1357–1372.

Motte, H., Vereecke, D., Geelen, D., & Werbrouck, S. 2014. "The molecular path to in vitro shoot regeneration." *Biotechnology Advances 32*(1): 107–121.

Murthy, H N., Sebastian, J K., Paek, K Y., & Park, S Y. 2023. "Bioreactor systems for micropropagation of plants: present scenario and future prospects." *Frontiers in Plant Science* 14: 1159588.

Nirmal, D., Teraiya, S., & Joshi, P. 2023. "Liquid culture system: an efficient approach for sustainable micropropagation." *Current Agriculture Research Journal 11*(1): 28–42.

Perez-de-Castro, A M., Vilanova, S., Cañizares, J., Pascual, L., Esteras, C., Diez, M J., Prohens, J., & Picó, B. 2012. "Application of genomic tools in plant breeding." *Current Genomics 13*(3): 179–195.

Potrykus, I. 1990. "Gene transfer to plants: assessment and perspectives." *Physiologia Plantarum 79*(1): 125–134.

Power, J. B., & Cocking, E. C. 1970. "Isolation of leaf protoplasts: macromolecule uptake and growth substance response." *Journal of Experimental Botany 21*(1): 64–70.

Prohens, J. 2011. "Plant breeding: a success story to be continued thanks to the advances in genomics." *Frontiers in Plant Science 2*: 51.

Reuther, G. 1990. "Current status and future prospects of large scale micropropagation in commercial plant production." *Food Biotechnology 4*(1): 445–459.

Sattiraju, K S., Kotiyal, S., Arora, A., & Maheshwari, M. 2018. "Plant growth-promoting microbes: contribution to stress management in plant hosts." In Sobti, R., Arora, N., Kothari, R. (eds) *Environmental Biotechnology: For Sustainable Future*. Springer, Singapore: 199–236. https://doi.org/10.1007/978-981-10-7284-0_8

Schoof, H., Lenhard, M., Haecker, A., Mayer, K. F., Jürgens, G., & Laux, T. 2000. "The stem cell population of Arabidopsis shoot meristems is maintained by a regulatory loop between the CLAVATA and WUSCHEL genes." *Cell 100*(6): 635–644.

Sharma, H. C., Crouch, J. H., Sharma, K. K., Seetharama, N., & Hash, C. T. 2002. "Applications of biotechnology for crop improvement: prospects and constraints." *Plant Science* 163(3): 381–395.

Sharma, N., Schneider-Canny, R., Chekhovskiy, K., Kwon, S., & Saha, M. C. 2020. "Opportunities for increased nitrogen use efficiency in wheat for forage use." *Plants 9*(12): 1738.

Sharma, S. K., & Millam, S. 2004. "Somatic embryogenesis in *Solanum tuberosum* L.: a histological examination of key developmental stages." *Plant Cell Reports 23*: 115–119.

Sihachakr, D., Haïcour, R., Cavalcante Alves, J. M., Umboh, I., Nzoghé, D., Servaes, A., & Ducreux, G. 1997. "Plant regeneration in sweet potato (*Ipomoea batatas* L., Convolvulaceae)." *Euphytica 96*: 143–152.

Smýkal, P., Nelson, M. N., Berger, J., & Wettberg, E. J. B. V. 2018. "The impact of genetic changes during crop domestication." *Agronomy 8*(7): 119.

Spindel, J., & McCouch, S R. 2016. "When more is better: how data sharing would accelerate genomic selection of crop plants." *New Phytologist 212*(4): 814–826.

Steingroewer, J., Bley, T., Georgiev, V., Ivanov, I., Lenk, F., Marchev, A., & Pavlov, A. 2013. "Bioprocessing of differentiated plant in vitro systems." *Engineering in Life Sciences 13*(1): 26–38.

Su, Y. H., Liu, Y. B., & Zhang, X. S. 2011. "Auxin–cytokinin interaction regulates meristem development." *Molecular Plant 4*(4): 616–625.

Tang, H., Sezen, U U., & Paterson, A H. 2010. "Domestication and plant genomes." *Current Opinion in Plant Biology 13*(2): 160–166.

Teixeira da Silva, J. A., & Jha, S. 2016. "Micropropagation and genetic transformation of *Tylophora indica* (Burm. f.) Merr.: a review." *Plant Cell Reports 35*: 2207–2225.

Vasil, I. K. 1994. "Molecular improvement of cereals." *Plant Molecular Biology 25*: 925–937.

Wójcik, A. M., Wójcikowska, B., & Gaj, M. D. 2020. "Current perspectives on the auxin-mediated genetic network that controls the induction of somatic embryogenesis in plants." *International Journal of Molecular Sciences 21*(4): 1333.

10 Future Horizons and Emerging Trends in Plant Regeneration Research

Pathakakula Srinivasu, Motha Kalpana, and Veluru Aparna

10.1 INTRODUCTION

Plant regeneration is a key biological process that enables plants to develop again by replacing or renewing aged, missing, or damaged parts (Bowman et al., 2000). For plant species to survive, adapt, and continue living in a variety of habitats, this complex and tightly controlled phenomenon is necessary (Sugimoto et al., 2019). The ability for regeneration highlights the extraordinary adaptability and tenacity of plant life, from the regeneration of injured organs and tissues to the regrowth of whole plants from individual cells. Cell division, tissue differentiation, and the restoration of functional organs are merely a few of the processes that are involved in plant regeneration (Ikeuchi et al., 2019). Rebuilding the structure and function of a plant is essentially facilitated by the activation of dormant developmental programmes that direct cells in that direction. In addition to being essential to a fundamental understanding of plant biology, knowledge of the processes regulating plant regeneration is also highly relevant to applied disciplines including biotechnology, horticulture, and agriculture (Raza et al., 2020).

Plant regeneration may occur in a variety of natural settings, such as in response to physical harm, herbivory, or environmental stress. Furthermore, many plants have special regeneration properties that allow them to respond to certain developmental stimuli, such as the emergence of adventitious roots or shoots (Raza et al., 2020). Using plant regeneration to its full capacity can lead to the production of genetically modified plants as well as the propagation of desired features, which have major implications for agriculture and agricultural improvement. Tissue culture, micropropagation, and genetic engineering are examples of biotechnology advances that have been made possible by improvements in plant regeneration methods. Owing to these technological advancements, horticulturists and scientists may now protect endangered species, grow plants on a massive scale, and integrate advantageous features into crops to increase agricultural production and resilience. Genetics, developmental biology, biotechnology, and environmental science all converge at the interface of plant regeneration research. In addition to expanding our knowledge of plant biology, deciphering the complex molecular processes and cellular dynamics driving plant regeneration presents opportunities for addressing global issues such

DOI: 10.1201/9781032702391-10

as sustainable agriculture, hunger security, and biodiversity preservation (Numan et al., 2021). Under these circumstances, research on plant regeneration is a lighthouse pointing to the deep complexity of life and the possibility of using nature's adaptability to our advantage.

10.2 HORIZONS IN PLANT REGENERATION RESEARCH

Plant regeneration has undergone a rapid evolution in recent years, driven by improvements in technology and a better comprehension of the molecular and cellular mechanisms governing plant growth and development. The amazing capacity of plants to regenerate and repair tissues, known as plant regeneration, has drawn much interest because of their significant applications in biotechnology, agriculture, and ecological restoration. Several noteworthy trends have developed as scientists probe further into the complex mechanisms underpinning plant regeneration, changing the course of this area, and opening new directions for both practical and scientific applications.

10.2.1 Micropropagation and Tissue Culture Inventions

Micropropagation and tissue culture techniques are being used to overcome challenges with plant regeneration related to efficiency and scalability. To improve the success rates of tissue culture-based regeneration methods, scientists are looking at new formulations for the culture medium, optimal growth environments, and additional ways that plants might regenerate – especially nonmodel species.

10.2.2 Epigenetics and Transgenerational Effects

Recent studies have revealed the impact of epigenetic modifications on plant regeneration. Researchers are examining how epigenetic changes, such as DNA methylation and histone modifications, affect the ability of plants to regenerate. Additionally, investigations into transgenerational effects are exploring how epigenetic changes acquired during regeneration might be inherited, offering a fascinating new perspective on plant adaptability.

10.2.3 Precision Genome Editing

CRISPR-Cas9 technology has substantially impacted plant regeneration research. This sophisticated technology enables precise genome editing in plants, allowing researchers to target specific genes involved in the regeneration pathway. As a result, it not only broadens our understanding of the genetic pathways underpinning regeneration, but also presents significant opportunities for producing crops with improved regenerative capacity and beneficial features.

10.2.4 Advanced Imaging and Osmics Technology

The molecular processes involved in plant regeneration are becoming increasingly clear because of the use of state-of-the-art omics tools including transcriptomics, proteomics, and genomics. Researchers have investigated the dynamics of the cells during regeneration thanks to advanced imaging methods. Combining these methods improves our understanding of the intricate systems that regulate the control of plant regeneration.

10.2.5 Synthetic Biology Approaches

Plant regeneration research increasingly includes synthetic biology. To precisely regulate regeneration processes, scientists are creating artificial gene circuits and pathways. This novel strategy broadens the potential uses for plants by facilitating the engineering of certain regeneration-related features and paving the way for the creation of synthetic tissues with predetermined functionalities.

10.2.6 Metabolic Engineering for Enhanced Regeneration

Optimising plant regeneration is mostly dependent on metabolic engineering. Researchers are attempting to increase the synthesis of important chemicals, hormones, and signalling components involved in regeneration by modifying metabolic pathways. This strategy not only increases the efficiency of regeneration procedures but also provides chances to customise plant regeneration for particular uses, such as the synthesis of bioactive chemicals.

10.2.7 Transcriptomics

Transcriptomics is a branch of molecular biology that involves the study of the complete set of RNA transcripts produced by the genome of an organism, known as the transcriptome. The introduction of CRISPR-Cas9 has greatly altered plant regeneration research. This sophisticated technique allows scientists to precisely modify plant genomes, allowing them to target specific genes involved in the regeneration process. As a result, this not only broadens our understanding of the genetic basis of regeneration, but it also holds significant promise for generating crops with increased regenerative power and desirable features.

10.3 EMERGING TRENDS IN PLANT REGENERATION RESEARCH

Plant regeneration research has a bright future ahead, owing to technological advancements, a growing understanding of plant biology, and the growing need for ecologically benign and sustainable agricultural practices. A few major horizons shape this research's path as we move forward.

10.3.1 Molecular Basis of Plant Regeneration

Recent progress has been made in understanding the cellular and molecular principles underpinning plant regeneration. Model plants such as *Arabidopsis thaliana* have been used for in-depth investigations of each of the aforementioned regeneration stages. Each phase, which governs cell dedifferentiation (somatic embryogenesis)

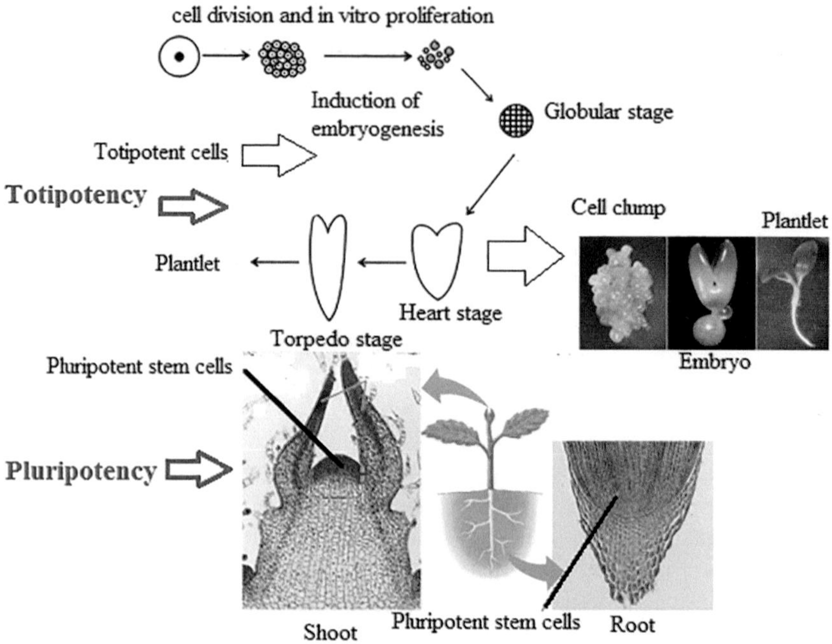

FIGURE 10.1 Plant regeneration processes. (Adapted from Bidabadi and Jain 2020.)

or reprogramming (*de novo* organogenesis), is governed by a group of molecular actors involved in developmental and signalling pathways (Chokheli et al., 2020). Plants have evolved the remarkable ability to regenerate tissues from organs that have undergone differentiation, a process that involves turning some cell types into others. Because of this flexibility, cell fates may be reprogrammed and cells dedifferentiate to regenerate whole organs and plants (Figure 10.1). Three principal classifications of regeneration: (1) tissue regeneration, (2) somatic embryogenesis, and (3) *de novo* organogenesis.

10.3.1.1 Tissue Regeneration

The basic mechanism of plant tissue regeneration enables plants to recover, adjust, and expand in response to a variety of stimuli. Plants have the innate ability to reprogram cells for specific activities, which drives the development of new tissues and organs. This is a complex biological phenomenon. The process of regeneration involves the start of organogenesis, also known as somatic embryogenesis, in which undifferentiated cells undergo a sequence of processes that result in the growth of roots, shoots, or embryos. Auxins, cytokinins, and other plant hormones are important for directing tissue development and coordinating cellular responses (Skoog, 1957). A frequent indicator of injury is the formation of a callus, which functions as a repository for pluripotent cells vital for tissue regeneration. The effectiveness of tissue regeneration is influenced by environmental variables such as temperature, light, and nutrient availability. Tissue culture techniques facilitate the widespread propagation of plants with specific traits and the development of genetically modified organisms. Moreover, progress in understanding the genetic control of plant regeneration has important implications for agriculture. In addition to enhancing our knowledge of plant development, research into plant tissue regeneration offers valuable insights into biotechnology and sustainable agricultural practices.

10.3.1.2 Somatic Embryogenesis

Somatic embryogenesis is a crucial biotechnological approach that offers significant advantages in clonal propagation and genetic transformation, among other applications. This process was first demonstrated in 1958 by Stewart, who successfully developed embryos in carrot embryos via suspension culture. Somatic embryogenesis is an artificial process in which somatic cells differentiate into somatic embryos, leading to the development of a plant or embryo from a single somatic cell. Unlike natural embryogenesis, this process involves the formation of embryos from plant cells that typically do not participate in embryo development. Somatic embryos do not develop a seed coat or endosperm around them (Bowman et al., 2000). The process begins with a single cell or a cluster of cells that initiate a developmental pathway, leading to the repeated regeneration of nonzygotic embryos capable of germinating into a complete plant. These cells are taken from selected source tissues and cultured in a medium to promote the growth of an undifferentiated mass of cells known as the callus. In the culture medium, plant growth regulators are used to induce callus formation, which can then be modified to stimulate the development of somatic embryos.

Developmental stages of SE:

1. Callus initiation
2. Embryo development and maturation
3. Plantlet formation

10.3.1.3 *De novo* Organogenesis

This approach enables the regeneration of complete organs that were not originally present. There are two types of *de novo* organogenesis: direct and indirect regeneration. Direct regeneration is the development of organs from explants without intermediate stages, whereas indirect regeneration includes an intermediate stage in which an undifferentiated callus is created. Some plants, such as succulents from the Cactaceae and Crassulaceae families, as well as *Jatropha curcas*, may grow new roots and shoots from stem cuttings (Severino et al., 2011). Plants, such as lettuce, undergo indirect organogenesis, which involves the formation of new branches from callus tissue.

10.3.1.3.1 Organogenic Callus Formation

Callus formation is the first phase in indirect organogenesis and is characterised by cell dedifferentiation into totipotent cells. This development can be triggered by a variety of stimuli, including auxin and injury, and is frequently accompanied by many gene expression pathways. Auxin-induced calli in Arabidopsis have molecular characteristics with pluripotent root tip cells and are derived from pluripotent pericycle cells at the xylem poles (Bull et al., 2020). Because their root cell structures are similar, the production of these auxin-induced calli closely parallels the growth of lateral roots. Wound-induced callus development, on the other hand, includes the activation of cytokinin signalling, which differs from the pathways involved in lateral root synthesis. Some researchers consider callus creation to involve trans-differentiation rather than dedifferentiation because of the parallel gene expression patterns between callus formation and other developmental processes.

10.3.1.3.2 De novo Root Organogenesis

De novo root organogenesis is the production of new roots from a plant's nonroot tissues or organs rather than existing roots. This process involves the establishment and growth of roots in response to certain signals or stimuli. It can be caused by a variety of factors, including auxins, mechanical stress, tissue damage, callus development, external signalling molecules, and genetic alterations (Bustillo-Avendano et al., 2018). Understanding the molecular and cellular mechanisms behind innovative root organogenesis is critical for applications such as stress tolerance enhancement, plant propagation improvement, and crop development with more efficient root systems for improved nutrient and water absorption. Researchers have studied and manipulated these processes to advance agricultural practices and improve ecological sustainability.

10.3.1.3.3 De novo Shoot Organogenesis

Shoot organogenesis can take place directly from an explant or indirectly through callus development (Figure 10.2). At the molecular level, callus cells are more similar to root tip cells than shoot cells are, necessitating changes in gene expression to shift them toward shoot growth rather than root growth. When shoots regenerate from calluses, a main meristem called the shoot apical meristem (SAM) is generated. The

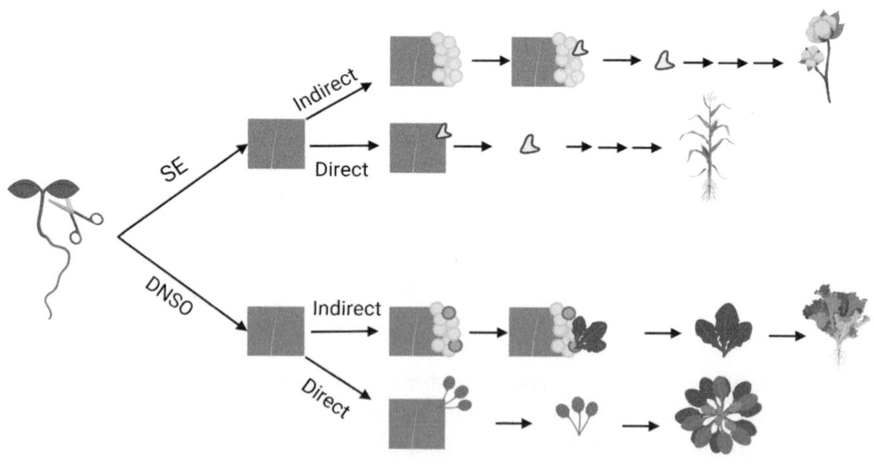

FIGURE 10.2 Pathways involved in the in vitro regeneration of vascular plants (somatic embryogenesis (SE) and *de novo* shoot organogenesis (DNSO)). (Adapted from Bull et al., 2020.)

SAM, like the root apical meristem (RAM), is made up of pluripotent stem cells that give birth to all of the plant's aboveground sections. The SAM's organising centre (OC) works similarly to the root apical meristem's quiescent centre (QC), preserving stem cell pluripotency via a feedback loop involving the WUSCHEL (WUS) protein and the signalling peptide CLAVATA3 (CLV3). The outer sections of the SAM produce leaves and lateral organs, whereas the stem is formed from basal cells in the rib zone. The SAM also has a core zone with a reservoir of stem cells that continually replace the peripheral and rib zones as they develop (Bull et al., 2020). The SAM contains large quantities of cytokinins, as opposed to the RAM, which is rich in auxin. The spatial distribution of auxin and cytokinin within these regions plays a crucial role in guiding the differentiation of pluripotent cells into either shoot or root cells.

Shoot regeneration from calli occurs in four stages:

1. Formation of a pluripotent callus
2. Shoot promeristem formation
3. Shoot progeneration
4. Shoot regeneration

10.4 GENOME EDITING TOOLS FOR PLANT REGENERATION

Recombinant DNA engineering has revolutionised the study of the genome to the next level to provide the opportunity for its application in various fields such as agriculture and industry. Genome editing is a technique in genetic engineering that involves inserting, deleting, or replacing DNA sequences within the genomes of living organisms via molecular tools known as nucleases. There are three primary families of nucleases used in this process:

1. Zinc finger nucleases (ZFNS)
2. Transcription activator-like effector-based effector nucleases (TALENs)
3. Clustered regularly interspaced short palindromic repeat (CRISPR)

10.4.1 Zinc Finger Nuclease (ZFN)

ZFNs are engineered restriction enzymes that modify or cut specific DNA targets using a ZFN DNA-binding domain (Zhang et al., 2010). These zinc finger domains can be custom-designed to bind to particular sequences in the host DNA. ZFNs consist of two main components: one is a DNA-binding domain constructed from eukaryotic transcription factors that includes a zinc finger and the other is a catalytic domain that contains the FokI nuclease, which cleaves the specific DNA sequences (Numan et al., 2021). ZFNs have been effective in elucidating the functions of various genes across different organisms, proving especially useful in identifying gene roles in diverse cellular contexts. However, ZFN technology faces challenges such as design complexity, high construction costs, and variable success rates.

10.4.2 Transcriptional Activator-Like Effector Nuclease (TALEN)

TALENs are engineered restriction enzymes designed to cut specific DNA sequences using TAL effector DNA-binding domains. Precise targeting is facilitated by a straightforward "code" that corresponds to the di-amino acid sequence (repeat-variable di-residue) within a conserved target sequence of approximately 33–35 amino acids (Numan et al., 2021). Advances in technologies for modifying genes and the establishment of methods for the streamlined synthesis and assembly of TALENs have made it possible to perform efficient editing at multiple sites (Figure 10.3).

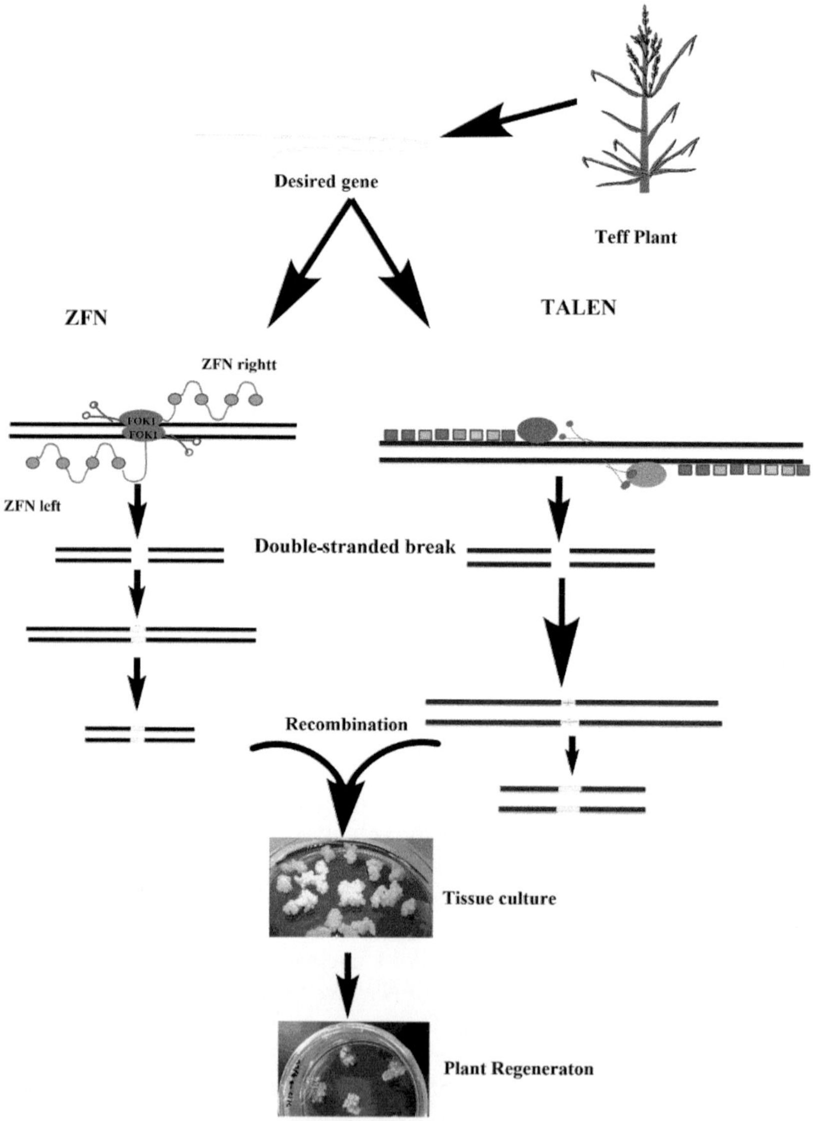

FIGURE 10.3 Schematic view of plant regeneration by zinc finger nuclease (ZFN) and transcriptional activator-like effector nuclease (TALEN). (Adapted from Numan et al., 2021.)

10.4.3 Clustered Regularly Interspaced Short Palindromic Repeat (CRISPR)

The CRISPR-Cas system is a prokaryotic immune mechanism that provides resistance against foreign genetic elements, such as those found in plasmids and phages, and serves as a form of acquired immunity. This system is the foundation of a genome editing technology called CRISPR-Cas, which enables permanent alterations of genes within organisms. The CRISPR-Cas system includes two primary components that facilitate changes in DNA sequences: Cas proteins and guide RNAs (gRNAs) (Numan et al., 2021). Often referred to as a third-generation programmable nuclease, the CRISPR-Cas system plays a significant role in crop protection. It is categorised into three main types (types I–III), which are further subdivided into 11 distinct subtypes. Each type is associated with specific Cas proteins, which are named on the basis of the model organism from which they are derived (Shelake et al., 2019).

10.4.3.1 CRISPR-Cas9 System

Cas9 is an RNA-guided DNA endonuclease that targets and inhibits foreign DNA. The gRNAs that direct Cas9 are derived from CRISPRs (Figure 10.4). CRISPRs are composed of tandem arrays of short, direct repeat sequences of 30–40 bp, separated by spacer sequences matching foreign DNA (Jiang et al., 2013). When CRISPR sequences are transcribed and processed, they produce mature CRISPR (cr)RNAs, which are flanked by signature CRISPR repeat tags at both the 5′ and 3′ ends. These (cr)RNAs form a complex with Cas proteins, creating a ribonucleoprotein (crRNP) that cleaves the invader's DNA/RNA. One of the remarkable features of CRISPR is its specificity, facilitated by the gRNA, which allows precise binding to target DNA. The system's versatility is enhanced by the ability to engineer custom gRNAs (Ran et al., 2013). Upon recognition of the target site by the gRNA, the Cas9 nuclease, aided by its RucV and HNH domains, cleaves the DNA strand, generating blunt-ended double-strand breaks (DSBs). These DSBs can be repaired by nonhomologous end joining (NHEJ), introducing mutations at the targeted site, or by homologous directed repair (HDR), which can knock in or replace desired gene fragments at the target site via template DNA.

10.5 REGENOMICS

Regenomics is a plant regeneration-dedicated database that contains all currently available transcriptomic data related to regeneration, specifically wound-induced tissue repair, hormone-induced plant regeneration, somatic embryogenesis, and protoplast regeneration, along with an analytical pipeline that processes user-submitted queries. The REGENOMICS analytical pipeline is divided into three analysis modes (single-gene analysis, multigene analysis, and transcriptome analysis) based on the kind of query (a single gene of interest (GOI), a list of GOIs, or a whole transcriptome). The single-gene analysis mode allows one to analyse the expression pattern, coexpression network constructed via bulk RNA-seq, and single-cell RNA-seq-derived GRNs of any GOI implicated in plant regeneration. Comparing expression patterns, building coexpression networks and GRNs, and performing differentially expressed gene

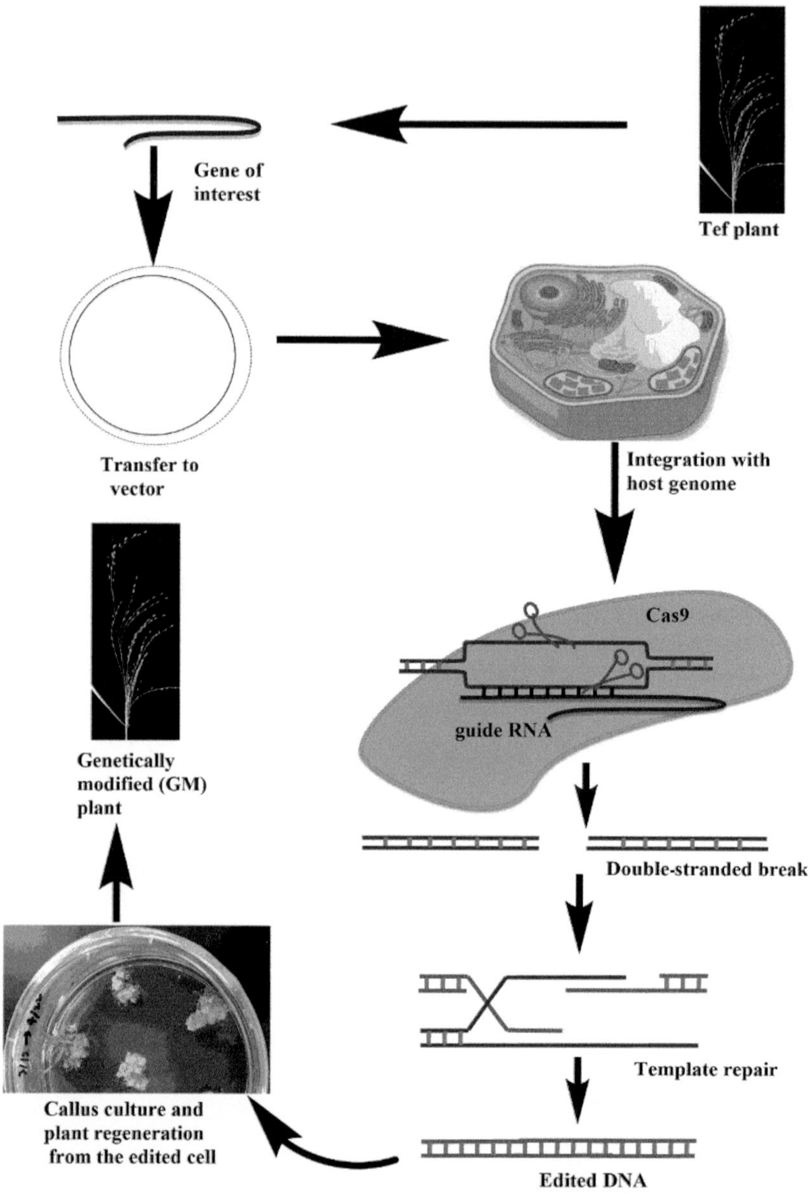

FIGURE 10.4 Schematic view of plant genome editing by CRISPR-Cas9. (Adapted from Numan et al., 2021.)

(DEG) enrichment analysis for a list of GOIs are all enabled by the multigene analysis mode. Using the transcriptome analysis mode, users can select transcriptomes from REGENOMICS's deposit and those they self-submitted for use in differential expression analysis, correlation analysis, and dimensionality reduction (Bae et al., 2022).

The regenomic application has five main features (Bae et al., 2022). First, searching for the gene name or gene locus code will make it simple to obtain the gene expression identities of any GOI(s) engaged in any method of plant regeneration. Second, there are five distinct coexpression networks in regenomics: one network was created from all bulk RNA-seq datasets deposited in the field and the other four networks were built from transcriptome datasets related to wound-induced tissue repair, hormone-induced plant regeneration, somatic embryogenesis, or protoplast regeneration. Third, DEG enrichment analysis allows the estimation of the biological and functional aspects of a group of genes. Fourth, the regenomics database contains single-cell transcriptome data of regenerated tissues as well as root and SAMs. Finally, users can input any Arabidopsis transcriptome profile (RNA-seq data) in comma-separated value (CSV) format to analyse its transcriptome-level correlation via the selected regenomics-deposited transcriptome data and user-deposited RNA-seq data.

10.5.1 Single-Gene Analysis

In single-gene analysis mode, the REGENOMICS web application supports three different analyses (gene expression browser, coexpression network, and single-cell analysis) using all currently available transcriptome data related to wound-induced tissue repair, hormone-induced plant regeneration, somatic embryogenesis, and protoplast regeneration. Users can retrieve information for any gene by searching for a gene name or the gene locus code. In the gene expression browser, the GOI expression values normalised in transcripts per million (TPM) are displayed as line plots or bar plots for each study under the selected mode of plant regeneration (Bae et al., 2022). To validate the gene expression browser of REGENOMICS, we queried many genes implicated in wound-induced tissue regeneration, hormone-induced plant regeneration, somatic embryogenesis, or protoplast regeneration.

10.5.2 Multigene Analysis

Multigene analysis allows the simultaneous analysis of multiple queries. The multigene analysis mode enables four different analyses (gene expression heatmap, coexpression network, DEG enrichment, and single-cell analysis) for each mode of plant regeneration simply by selecting the menu tab and then submitting a list of genes by their gene locus codes. For gene expression heatmap analysis, users submit multiple query genes and select a transcriptome dataset (a mode of plant regeneration) for heatmap visualisation. Rows are ordered by hierarchical clustering on the basis of gene expression patterns, and subsets of the dendrogram can be magnified for a closer look. In this study, as an example, we queried CUC1, CUC2, ESR1/DRN, ESR2/DRNL, PLTs, STM, WOX5, and WUS, key regulators of hormone-induced plant regeneration, and showed that their expression tendencies may be compared throughout the process of hormone-induced plant regeneration (Bae et al., 2022).

10.5.3 Transcriptomics

Transcriptomics is an advanced discipline of molecular biology that focuses on the analysis of all RNA molecules found in a cell or tissue at a given time. In the context of plant regeneration, transcriptomics is an essential tool for comprehending the complex molecular mechanisms involved in tissue culture and plants. By analysing the transcriptome, or the entire set of transcripts in a cell, researchers may identify and understand the genes that are repressed or activated during different stages of plant regeneration. This information is essential for improving tissue culture methods, increasing plant regeneration success rates, and improving the efficacy of regeneration processes (Ikeuchi et al., 2019). With the application of transcriptomics, researchers may obtain important knowledge on the genetic control of cellular processes, hormone signalling pathways, and stress responses related to plant regeneration. This knowledge can be applied to the biotechnological development of robust and productive plant breeds.

10.6 PROTEOMICS

Understanding the dynamic protein interactions and expression during plant regeneration is crucial, and proteomics, the complete analysis of proteins produced within a biological system, is important in this context. Researchers may discover, measure, and characterise the particular proteins involved in crucial phases of plant regeneration through the use of proteomic approaches. Plant regeneration is a diverse process that involves both molecular and cellular processes. The regulatory networks, signalling pathways, and protein changes that control cellular responses during tissue culture and regeneration can be better understood by examining the proteome of plants (Ikeuchi et al., 2019). Plant regeneration procedures may be optimised, the total efficiency of plant regeneration systems can be increased, and the molecular subtleties of plant regeneration can be understood on the basis of this knowledge. Proteomics makes a substantial contribution to the advancement of our understanding of the underlying molecular mechanisms, which in turn helps to increase the sustainability and effectiveness of plant regeneration for use in biotechnology and agriculture.

10.7 SYNTHETIC BIOLOGY

Synthetic biology provides a revolutionary approach to plant tissue regeneration by using engineering concepts to produce systems of biology that are more capable of regeneration. Synthetic biology makes it possible to precisely alter plant genomes to modify important genetic and molecular processes related to tissue regeneration. Researchers have the ability to engineer plants in such a way that they respond to regeneration more quickly or more effectively, which would promote faster tissue development and repair (Cachat et al., 2011). This might entail the optimisation of elements controlling callus development, the insertion of synthetic genetic circuits to regulate cellular processes, or the targeted manipulation of hormone signalling pathways. With the use of synthetic biology techniques, scientists may also modify plants to better withstand and adapt to certain environmental circumstances. The application of synthetic biology concepts to the study of plant tissue regeneration might result in the development of crops that are more efficient, resilient to stress, and able to coexist in a variety of ecological niches, all of which would promote sustainable agriculture.

10.8 CONCLUSION

In conclusion, the future of plant regeneration research involves a convergence of cutting-edge technologies, interdisciplinary collaborations, and a deeper understanding of plant biology. The integration of these emerging trends holds the promise of transforming agriculture, ensuring food security, and contributing to a more sustainable and resilient global ecosystem. As researchers explore these frontiers, the potential benefits of addressing global concerns in agricultural and environmental sustainability are both substantial and inspiring.

REFERENCES

Bae, S. H., Noh, Y. S., and Seo, P. J., 2022. "A web-based application for plant regeneration-associated transcriptomics analyses." *Computational and Structural Biotechnology Journal* 2: 3234–3247. https://doi.org/10.1016/j.csbj.2022.06.033

Bidabadi, S. S., and Jain, S. M. 2020. "Cellular, molecular, and physiological aspects of in vitro plant regeneration." *Plants* 9 (6): 702. https://doi.org/10.3390/plants9060702

Bowman, J. L., and Eshed, Y., 2000. "Formation and maintenance of the shoot apical meristem." *Trends Plant Science* 5 (3): 110–115. https://doi.org/10.1016/S1360-1385(00)01569-7

Bull, T., and Michelmore, R. 2020 "Molecular determinants of in vitro plant regeneration: Prospects for enhanced manipulation of lettuce (*Lactuca sativa* L.)." *Frontiers in Plant Science* 13: 888425. https://doi.org/10.3389/fpls.2022.888425

Bustillo-Avendaño, E., Ibáñez, S., Sanz, O., Sousa Barros, J. A., Gude, I., Perianez-Rodriguez, J., Micol, J. L., Del Pozo, J. C., Moreno-Risueno, M. A., and Pérez-Pérez, J. M. 2018. "Regulation of hormonal control, cell reprogramming, and patterning during *de novo* root organogenesis." *Plant Physiology* 176 (2): 1709–1727. https://doi.org/10.1104/pp.17.00980

Cachat, E., and Davies, J. 2011. "Application of synthetic biology to regenerative medicine." *Bioengineering and Biomedical Science* 2: 003. https://doi.org/10.4172/2155-9538.S2-003

Chokheli, V. A., Dmitriev, P. A., Rajput, V. D., Bakulin, S. D., Azarov, A. S., Varduni, T. V., Stepanenko, V. V., Tarigholizadeh, S., Singh, R. K., Verma, K. K., and Minkina, T. M. 2020. "Recent development in micropropagation techniques for rare plant species." *Plants* 9 (12): 1725–1733. https://doi.org/10.3390/plants9121733

Ikeuchi, M., Favero, D. S., Sakamoto, Y., Iwase, A., Coleman, D., Rymen, B., and Sugimoto, K. 2019. "Molecular mechanisms of plant regeneration." *Annual Review of Plant Biology* 70 (1): 377–406. https://doi.org/10.1146/annurev-arplant-050718-100434

Jiang, W., Zhou, H., Bi, H., Fromm, M., Yang, B. and Weeks, D.P., 2013. "Demonstration of CRISPR/Cas9/sgRNA-mediated targeted gene modification in Arabidopsis, tobacco, sorghum and rice." *Nucleic Acids Research* 41 (20): 188. https://doi.org/10.1093/nar/gkt780

Numan, M., Khan, A. L., Asaf, S., Salehin, M., Beyene, G., Tadele, Z., and Ligaba-Osena, A., 2021. "From traditional breeding to genome editing for boosting productivity of the ancient grain tef [*Eragrostis tef* (Zucc.) Trotter]." *Plants* 10 (4): 628. https://doi.org/10.3390/plants10040628

Ran, F. A., Hsu, P. D., Wright, J., Agarwala, V., Scott, D. A., and Zhang, F., 2013. "Genome engineering using the CRISPR-Cas9 system." *Nature Protocols* 8 (11): 2281–2308. https://doi.org/10.1038/nprot.2013.143

Raza, G., Singh, M. B., and Bhalla, P. L., 2020. "Somatic embryogenesis and plant regeneration from commercial Soybean cultivars." *Plants* 9 (1): 38. https://doi.org/10.3390/plants9010038

Severino, L. S., Lima, R. L., Lucena, A. M., Freire, M. A., Sampaio, L. R., Veras, R. P., Medeiros, K. A., Sofiatti, V., and Arriel, N. H., 2011. "Propagation by stem cuttings and root system structure of *Jatropha curcas*." *Biomass and Bioenergy* 35 (7): 3160–3166. https://doi.org/10.1016/j.biombioe.2011.04.031

Shelake, R. M., Pramanik, D., and Kim, J. Y., 2019. "Exploration of plant-microbe interactions for sustainable agriculture in CRISPR era." *Microorganisms*, 7 (8): 269. https://doi.org/10.3390/microorganisms7080269

Skoog, F., 1957. "Chemical regulation of growth and organ formation in plant tissue cultures in vitro." *Symposia of the Society for Experimental Biology* 11: 118–131.

Sugimoto, K., Temman, H., Kadokura, S., and Matsunaga, S. 2019. "To regenerate or not to regenerate: Factors that drive plant regeneration." *Current Opinion in Plant Biology* 47: 138–150. https://doi.org/10.1016/j.pbi.2018.12.002

Zhang, F., Maeder, M. L., Unger-Wallace, E., Hoshaw, J. P., Reyon, D., Christian, M., Li, X., Pierick, C. J., Dobbs, D., Peterson, T., and Joung, J. K., 2010. "High frequency targeted mutagenesis in *Arabidopsis thaliana* using zinc finger nucleases." *Proceedings of the National Academy of Sciences of the United States of America* 107 (26): 12028–12033. https://doi.org/10.1073/pnas.0914991107

Index

Note: **Bold** page numbers refer to tables and *italic* page numbers refer to figures.